The Biggle Poultry Book

A CONCISE AND PRACTICAL

Treatise

ON THE

Management of Farm Poultry

BY

JACOB BIGGLE

ILLUSTRATED

"*What this country needs is less hog and hominy and more chicken and celery.*"

Skyhorse Publishing

TIM'S TWINS FIND THE STOLEN NEST.

Skyhorse Publishing books may be purchased in bulk at special discounts for sales promotion, corporate gifts, fund-raising, or educational purposes. Special editions can also be created to specifications. For details, contact the Special Sales Department, Skyhorse Publishing, 307 West 36th Street, 11th Floor, New York, NY 10018 or info@skyhorsepublishing.com.

Visit our website at www.skyhorsepublishing.com.

10 9 8 7 6 5 4 3 2 1

Library of Congress Cataloging-in-Publication Data is available on file.
ISBN: 978-1-62636-147-8

Printed in China

CONTENTS.

LIST OF COLORED PLATES.

Chapter I.

PRELIMINARY PARLEY.

This little book is intended to help farmers and villagers conduct the poultry business with pleasure and profit. Its teachings are not drawn from the author's inner consciousness exclusively, but from practical experience, study and observation.

I have been successful in the business myself, not as a fancier, but as a farmer, a fact which I do not attribute to my own ability entirely, but partly to the help derived from the stimulating and restraining influence of my good wife Harriet, and to Martha, the industrious and vigilant spouse of our faithful Tim.

A good deal of what I know and have written has really been derived from a diligent perusal of the *Farm Journal*, and I confess to having borrowed considerably from its pages both in text and illustration. Credit must therefore be given in a comprehensive way to the Poultry Editor of that publication, whose discerning mind and great experience with poultry have received the widest recognition by all interested in the poultry industry. I could do nothing better than to draw largely upon him, augmenting his practical information with trimmings from my own observation and experience, and with suggestions from the women folks and from Tim.

Great pains have been taken with the illustrations, and those having charge of this feature of the book deserve much praise for the skill, taste and originality displayed. They certainly have done well. The beautiful and life-like pictures set off the book in fine style and raise it far above the level of the commonplace.

The paintings for the colored prints were made from life from birds in the yards of breeders or on exhibition at the poultry shows, by Louis P. Graham, a young Philadelphia artist possessing a high order of talent. They are as true to nature and the ideal bird as it is possible to make them.

Few people have an adequate idea of the importance of the poultry business in this country. It is estimated that there are in the United States over three hundred millions of chickens and thirty millions of other domestic fowls. There are produced in one year nearly one billion dozen eggs of an average worth of ten cents per dozen, making the annual value of the total egg product one hundred million dollars. If in addition to this the yearly product of poultry meat is considered, the importance of this branch of rural economy will be more fully appreciated.

A pound of eggs or a pound of poultry can be

raised as cheaply as a pound of beef or mutton. Poultry sells at home for nearly twice the price per pound you get for beef and mutton on the hoof. Eggs sell for more than twice the price per pound on the farm that the city butcher gets for the dressed carcasses of the animals he sells.

I have not written this book for the poultry fancier, although that valued person will find many points of interest in it, but for the practical farm or village man or woman who raises poultry and eggs for market, whose flock is one of the many sources by which the income of the farm or village acre is increased with but a trifling money outlay, and with but little extra care and work. As in every other branch of farm production, however, poultry always responds quickly to any extra effort and thought put into it, and there are hundreds of farms to-day where the poultry yard yields more ready cash than any other department.

This book is small in measure; I could have doubled the size easily, but it would have been thinner and not any better, at least so it seems to me, and Harriet agrees. Should this be your verdict, gentle reader, I shall be content.

JACOB BIGGLE.

Elmwood, 1895.

PARTS OF THE CHICKEN.

1. Comb.
2. Face.
3. Wattles.
4. Ear-lobes.
5. Hackle.
6. Breast.
7. Back.
8. Saddle.
9. Saddle-feathers.
10. Sickles.
11. Tail-coverts.
12. Main tail-feathers.
13. Wing-bow.
14. Wing-coverts, forming wing-bar.
15. Secondaries, wing-bay.
16. Primaries or flight-feathers ; wing-butts.
17. Point of breast bone.
18. Thighs.
19. Hocks.
20. Shanks or legs.
21. Spur.
22. Toes or claws.

TYPES OF HEADS AND COMBS.

1. Single comb.
2. Spiked comb.
3. Rose comb.
4. Pea comb.
5. Cup comb.
6. Leaf comb.
7. Single comb, female.

PLATE I.

BARRED PLYMOUTH ROCKS.

CHAPTER II.

THE EGG.

Don't put all your eggs in one basket.—Old Proverb.
Put all your eggs in one basket, and watch that basket.
—Mark Twain's Version.

Careful and critical examination of an egg reveals an arrangement of its contents in a series of layers as seen in the illustration.

Referring to the cut, A is the shell; B is the membrane adhering to the shell; C is a second membrane slightly adhering to B, except at the large end, where the two separate and form D, the air space; E is the first layer of the white or albuminous part and is in liquid form; F is the second layer, which is semi-liquid, and G is the inner layer; H, H are the chalazæ, or slightly thickened membranes that unite the white to the membrane enclosing the yolk, M. They form a ligament that binds the parts together, and holds the yolk suspended in the midst of the white or albumen. I, J, K are very fine membranes surrounding the yolk; L is the germ, and N is the germ sack or utricle; a, b, c are separate layers composing the yolk. The germ, L, and germ sack, N, are suspended by the mem-

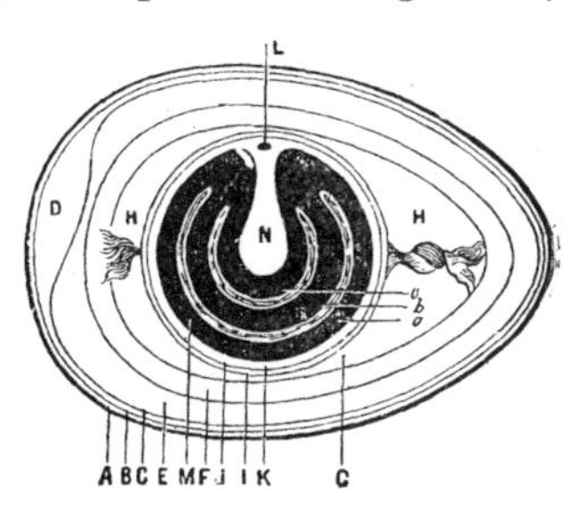

branes H, like a mariner's compass, so that the germ always retains its position on top of the yolk. While this germ is present in all eggs alike, it requires the contact of the male element to give it vitality. This contact takes place in the oviduct before the yolk is surrounded by the white, or albumen, and the shell.

The yolk is the essential part of the egg, containing as it does the germ, and albuminous and fatty matter and organic salts sufficient to support the germ in its earlier stages of development. The white, which is pure albumen and water, furnishes in the first place a safe and congenial medium for the preservation of the life germ and afterwards contributes its share of nutriment to the developing embryo.

The shell is a layer of carbonate of lime deposited so as to give the greatest possible strength, and so arranged as to leave numerous pores through which the water of the egg can escape and the external air can enter.

About three-fourths, 74 per cent., of the contents of an egg consist of water, 14 per cent. is albumen, 10.5 per cent. is fat, and 1.5 per cent. is ash. Of the latter the principal part consists of phosphate of lime, the element that enters so largely into the composition of bones.

These constituents of an egg furnish every element, except oxygen, essential to the formation of the living bird.

The egg is the beginning of all animal life. In the case of mammals, this egg is hatched and the young animal is nourished and developed for a certain period within the body of the mother before it is cast upon the cold charities of the world. The egg of a bird, or a reptile, is expelled as soon as it is perfectly formed, and the germ of life within it is awakened or destroyed by surrounding conditions.

The application of heat, 100 degrees to 103 degrees Fahrenheit, to the egg of the domestic fowl will cause the germ within to begin a process of transformation. Within twenty-four hours after incubation begins, an examination will show a zone of small blood vessels formed around this germ. After three days a temporary membrane begins to form inside of the shell membranes. This new membrane serves as lungs to the growing embryo; into its numerous hair-like vessels the contents of the egg are absorbed and changed into blood. This blood is exposed to the oxygen of the air that enters through the pores of the shell, and thus, purified and vitalized, returns to the centre of life, circulation is established and development proceeds rapidly until the entire egg is absorbed and transformed into a creature having various organs and a conscious life.

The different stages in the process of development above described, may be observed by breaking eggs that have been exposed for different periods to the proper conditions for incubation.

The contents should be turned out into a saucer, great care being taken not to rupture the delicate membranes that are forming. A good hand reading glass will greatly aid in making this examination.

As breaking the egg destroys the embryo, this method of examination is useful only to train the eye and judgment of the observer to examine the embryo through the shell. This may be done by holding the egg between the eye and a strong light. Various contrivances are used to assist the eye. One of the most simple, is made like a tin horn having a piece of soft leather or rubber over the large end and a hole in it, oval in shape, and a little smaller than the eggs to be tested. Such a tester may be made of tin or card board.

To test an egg, grasp it between the thumb and finger of the left hand and holding it large end up against the aperture of the tester look directly through it toward the light. While doing so revolve it slowly to get a view from all sides and to observe the motion of the embryo.

Figure 1 illustrates a tester that any handy person can make. The box is six inches square by eighteen inches high, open at top with a sliding door on one side. This holds a lamp. Opposite the lamp flame is a hole one and a half inches in diameter and around this a washer cut from a rubber boot. Back of the lamp place a piece of looking glass, and paint the rest of the box inside a dull black.

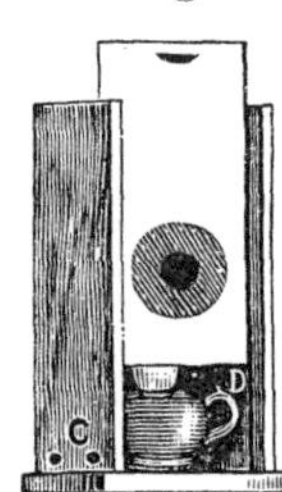

FIG. 1.

Have holes at bottom of box to ventilate lamp.

A fresh egg looks like Figure 2, almost perfectly clear. With a strong light and a thin white-shelled egg the outline of the yolk can be seen. Eggs with thick brown shells are difficult to test.

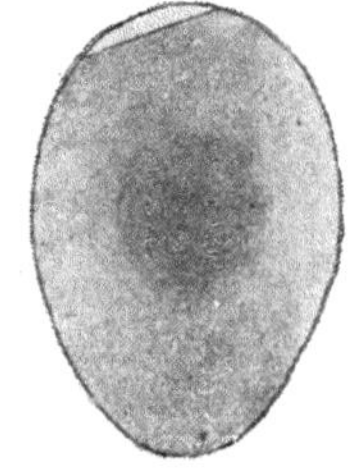
FIG. 2.

On the fifth or sixth day of incubation, a strong, fertile egg will look like Figure 3. The air-sack is slightly enlarged and from a dark center fine red lines are seen to radiate. There is also a slight cloudiness about this dark spot or germ, and the germ can be seen to move slightly as the egg is revolved.

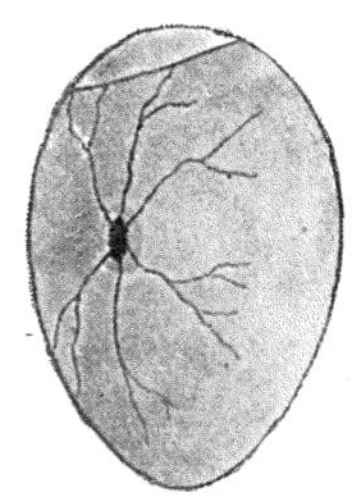
FIG. 3.

It often happens that the germ begins to develop and dies before the sixth day. In this case the red lines are indistinct, or absent, and in their place is a dark circle enclosing the germ as appears in Figure 4. When the egg is revolved this dead embryo floats aimlessly about in the surrounding contents.

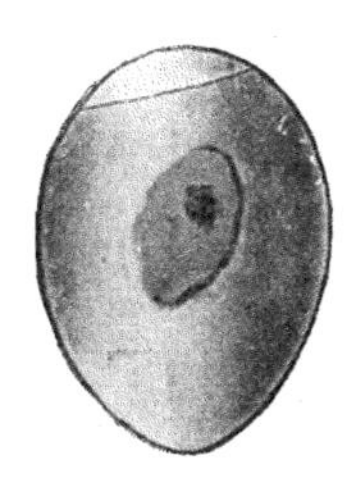
FIG. 4.

All infertile eggs that were fresh when incubation began, will remain clear up to the sixth day, or even longer, but a stale egg shows a cloudy spot in the center and a large air sack. When opened, the yolk sack is apt to break and the contents to run together, or, as we say, become "addled."

All such eggs, as well as those that contain dead embryos, and all clear or infertile eggs should be removed at this first testing.

A second testing of eggs should be made on the tenth day. By this time the air sack has still further enlarged and the growth of the embryo

FIG. 5.

has so clouded the egg contents as to render the outlines indistinct. The appearance of the egg is now shown by Figure 5.

After the tenth day the tester is of little use. On the eighteenth day the embryo is nearing the final stages, the yolk upon which it subsists is nearly all absorbed. On the nineteenth and twentieth days it is chipping the shell, and on the twenty-first it emerges, fully developed, into a new and larger world.

FOOT NOTES.

The shell of an egg is porous and any filth on it will taint the meat. A good reason for cleaning eggs as soon as gathered.

Sometimes dirty looking eggs are fresher than some that are clean, but buyers will not believe it, and, as they must judge an egg by its outward appearance only, eggs should be made as attractive looking as possible before being sent to market.

Eggs are preserved in two ways: By cold storage in a dry atmosphere, at a temperature of 36 to 40 degrees, and by immersing in a pickle of lime and salt in clean oak barrels. The pickle is made by slaking two pounds of lime in hot water, and adding one pint of salt and four gallons of water. Twenty gallons will cover 150 dozens. Put fresh eggs in the clear pickle until the vessel is nearly full, spread a clean cloth over them and cover this with the settlings of the lime.

Ice-house eggs and pickled eggs are edible if put in fresh and properly kept, but are greatly inferior to fresh stock. If sold for what they are it is all right, but it is all wrong and a fraud on consumers to palm them off as newly-laid eggs.

PLATE II.

SILVER LACED WYANDOTTES.

Chapter III.

EGGS FOR HATCHING.

To me eggs are like morals—they have no middle ground. If not good, they are bad.—Harriet.

O. W. Holmes is credited with the observation that a child's education should begin one hundred years before it is born. In this witticism the poet and sage expresses his appreciation of the law of heredity, that like begets like, a principle as applicable to the raising of fowls as to the training of children.

The successful chicken rearer must begin his operations long before the advent of the chickens. Hens that have been stunted by neglect and abuse or debilitated by too frequent intermingling of blood, will not lay eggs containing strong, healthy germs. The breeding birds of both sexes should be of hardy stock, fully matured and in a high state of health.

Young pullets forced into early laying by stimulating food do not make good breeders. Hens that are over two years old, hens that are over fat, or have been weakened by disease, should never be used to furnish eggs for hatching. Pullets that have reached their full size, and well preserved two-year-old hens mated with a vigorous male, make the best breeders. A good plan is to mate hens with a cockerel from eight to twelve months old, and to mate pullets with an active cock not over two years old. The exact age when a bird reaches maturity cannot be given, as the different breeds vary greatly in this respect.

In order to obtain eggs with germs of strong vitality, the diet of the breeders must receive attention. Eggs are produced from what we may call surplus food, that which is not required for the sustenance of the hen herself. As we have already seen, the egg contains substances that make fat, lean meat or muscle and bones. To reproduce these in eggs the hen must eat and digest substances out of which these are made. Starchy foods contain the necessary oil or fatty matter. These are represented by the grains, especially corn, wheat, buckwheat and barley, and vegetables, especially potatoes and sugar beets. The mineral element that is found in eggs is found also in nearly all foods. Of the grains, oats have the largest percentage, then follow barley, sweet corn, buckwheat and rye, wheat and corn in the order named. Wheat, bran, clover hay, linseed and cottonseed meal and buttermilk are all rich in this element. Of the twenty-six per cent. of solids in an egg, fourteen consist of albumen, from which may be seen the absolute necessity of supplying the laying hen with food containing a large proportion of albuminous matter. The alchemy of nature working in the body of the hen cannot elaborate albumen out of starch or fat, nor out of carbonate and phosphate of lime. Food abounding in these will not enable the hen to produce eggs, if it be deficient in what are called albuminoids or nitrogenous elements. While the grains contain these they are not contained in sufficient quantity to form a proper diet for egg production when the grains are fed alone. Resort is had, therefore, to foods rich in albuminoids. Meat-meal, made from lean meat dried and ground, is the richest in this respect of all the foods found in the market. After meat-meal, follow in order green cut bone, cottonseed meal, linseed meal, wheat bran, clover hay and milk.

The hens when running at large in the warm season of the year supplement the ration of grain supplied them by their keeper with worms, grubs and insects of various kinds, which contain the needful

HE FINDS A WORM.

albumen. While providing themselves with this they obtain succulent and bulky green food in the form of grass, and gritty particles to grind the whole mass.

Along with the needful quantity and variety of food, hens roaming the fields secure the exercise so essential to good health and the production of healthy progeny.

Eggs of strong vitality for hatching may be obtained even from hens in confinement when the conditions noted here are complied with.

The same conditions that promote health and induce the hens to lay are favorable for giving vigor to the cock also.

It is difficult to lay down definite rules in regard to the number of hens to be allowed for each male bird. Breeds and individuals of each breed differ in activity and vigor; but speaking generally, it may be said that for a flock at liberty, one Leghorn male may be allowed for each flock of twenty to twenty-five females; one Plymouth Rock male to fifteen to twenty females; and one Brahma male to ten to fifteen females; these breeds being taken to represent the small, medium and large fowls. When confined in yards, reduce the number of females by a third, unless two males are allowed each pen, alternated weekly. Never have more than one male with the flock at the same time.

To be sure that eggs for hatching are fertile, none should be saved for this purpose from a flock until the third day after mating.

After mating, though the male be removed, the eggs laid from the third to the tenth day will nearly all be fertile. It follows from this, that in breeding pure-bred fowls, contamination of the blood from the introduction of a strange male need not be feared after the tenth day.

Never shake an egg designed for hatching.

Wrap eggs kept for hatching in old flannel or woolen cloth, or stand on end in bran and cover with flannel. Avoid a hot, drying atmosphere.

Beware of breeding from cocks with crooked breasts, wry tails, long, slender shanks, or any other bodily defect indicating a lack of vigor. Like begets like. Use only the best for stock birds.

PLATE III.

LIGHT BRAHMAS.

Chapter IV.

HATCHING THE EGGS.

Eggs are close things, but the chicks come out at last.
—Chinese Proverb

Incubation is the application of the proper amount of heat to the egg under proper conditions. Nature has provided for this by bringing upon hens after laying a certain number of eggs, the brooding fever, which runs its course when its purpose has been fulfilled.

In some breeds this broody instinct has been bred out to a great extent. This is true of the smaller, or Spanish breeds generally, yet even these will occasionally become broody. Nearly all the medium sized breeds, and the larger ones, too, are persistent sitters. Of all the standard breeds, perhaps the Cochins are by nature the most quiet and gentle, and have the motherly instinct the most strongly developed.

Whatever may be the breed, it is best, as a rule, to select for sitters and mothers, medium sized hens, and such as are not too fat and clumsy. It is an advantage, also, to have those that are gentle and will not fidget and fight and break their eggs. Wild, squalling hens are a nuisance; accustom them to being handled, remove them at night to a room apart from the laying hens, let them sit for a day or two on nest eggs, and if they promise well, give them as many as they can cover well.

No invariable rule can be laid down respecting the number of eggs to be put under a hen. The size of the hen, the size of the eggs and the season of the year will determine the proper number, which may be from nine to eighteen.

The manner of making the nest, a very simple operation, apparently, has much to do with the success or failure of a hatch. The box in which the nest is made should be so large as not to prevent the hen from turning about freely, and so situated that she cannot be interfered with by other hens. One of the cheapest and most satisfactory nest boxes for general purposes is illustrated herewith. It is a large soap box with two-thirds of the top removed, turned on its side. A box of this kind set on the floor of the laying room or on a shelf with the open side toward the wall but a few feet from it, makes a handy and secluded nesting place. When a hen becomes broody, the box can be moved near the wall and other hens shut out, and at the proper time she can be carried on her own nest to the hatching-room.

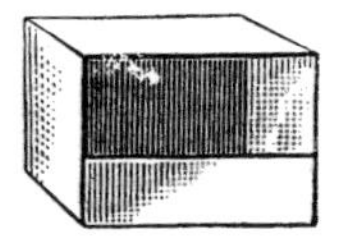
FIG. 1.

If a new nest must be made it should be of some soft material, broken oat straw or hay, carefully spread out and pressed down, hollowed but slightly, and the edges raised a little to prevent the eggs from rolling out. If the bottom be made too flat the eggs roll away from the hen and she cannot cover them ; if too convex, they roll close together, and when the hen enters the nest and steps on them or among them they do not separate or roll away and a fouled nest is the result. Whenever eggs are thus smeared or fouled in any manner, they should be carefully washed in warm water and at once replaced under the hen.

In selecting eggs for hatching, such as are very large or very small, all having unusually thin, rough or chalky shells, should be discarded.

It is a good plan to mark on every egg with pen and ink the date of sitting, and when they are due to hatch, and to make a record of the same in a book kept for the purpose. Always put the eggs under the hen after dark, unless she is known to be perfectly gentle and trustworthy.

To save labor it is a common custom to set several hens at one time, and when the chicks hatch to put two or more broods with one mother.

About the best food for sitting hens is corn. With corn, water, gravel, and a place to dust supplied, they will need little else. Their attendant should see that they come off the nest once a day and that their eggs are not fouled or broken.

The modern man-made hatcher, the incubator, is largely used for winter hatching when hens rarely become broody, and also for hatching on a larger scale than is convenient with the natural mother.

TYPE OF HOT-AIR INCUBATOR.

While the names and makers of these machines are numerous they are divided into two general classes, those warmed by hot air, and those warmed by radiation from a tank of hot water, the heat being supplied in both cases by a lamp flame or a gas jet. A very few are still made that are heated by drawing off the cooled water from a tank and pouring in hot water as required.

Each kind and each make has its friends, and nearly all are fairly successful. An expert having knowledge and experience in artificial hatching can make a success of the crudest incubator, while a person ignorant in such matters may fail with the most improved.

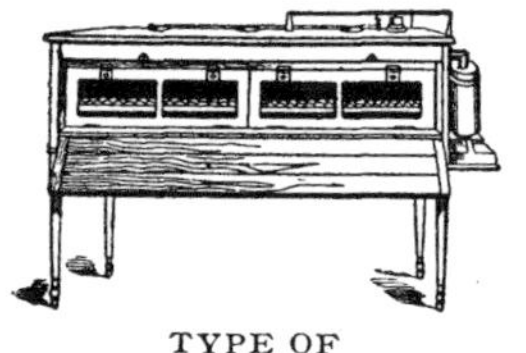

TYPE OF
HOT-WATER INCUBATOR.

The running of an incubator with only a few eggs in it at first, to learn how to manage it and to gain experience, is the part of wisdom for a novice. The directions sent by all manufacturers with their machines should be carefully studied during these experimental hatches.

The best location for an incubator is in a room where a mild and fairly uniform temperature can be preserved in spite of changes in the weather. Such a location is afforded by a light, dry and well ventilated basement or cellar. The machine should stand on a firm foundation, and where the direct rays of the sun cannot shine upon it.

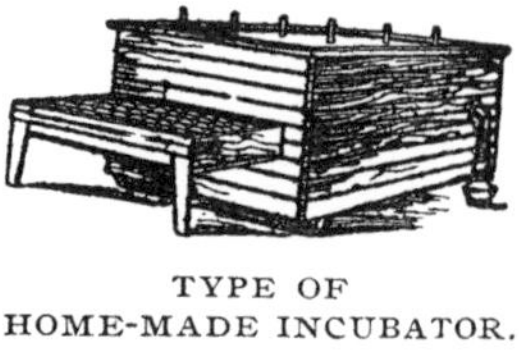

TYPE OF
HOME-MADE INCUBATOR.

Before filling the trays with eggs run it empty for a day or two to see that it is in working order, and that the heat can be maintained at 102 degrees to 104 degrees Fahrenheit.

Eggs for incubator hatching should be fresh, the fresher the better. None should be over ten days old, although they will hatch when much older if carefully preserved under woolen covers, and turned daily. The trays should be crowded at first, since, on testing the eggs on the fifth day, many may be found infertile and will have to be taken out.

After an incubator full of eggs has once been started, no additional eggs should be put in until the hatching is completed. This may be accepted as a rule to tie to without giving all the reasons for it here.

Eggs to hatch well must lose a part of the water contained in them. This loss occurs by evaporation through the pores of the egg-shell. Under the hen evaporation is checked just at the right time by a slight film of oil from the hen's body that shows itself in the gloss that appears on eggs that have been in the nest for a few days. In the incubator the evaporation will continue for the whole period of incubation and be excessive unless checked by supplying a moist atmosphere to the egg trays. Each manufacturer has his own method for furnishing the required moisture, and nearly all furnish moisture gauges or hygrometers for recording the amount of humidity in the egg chamber.

A reliable thermometer is one of the first essentials to success in artificial hatching. The secret of many failures may be traced to thermometers with scales inaccurately marked between the points 100 degrees and 105 degrees, just where accuracy is especially required in hatching eggs.

The proper temperature for hatching is considered to be 102 degrees to 103 degrees. This is the temperature, not of the egg chamber, but the temperature of the upper surface of a fertile, live egg. The temperature of an infertile egg, or of an egg containing a dead embryo will be lower than that of a live egg lying adjacent in the same tray. It is important, therefore, in testing the temperature to place the bulb upon a live egg.

By the tenth day the animal heat that has been stored in the living embryos in the process of incubation becomes quite a factor in the temperature of the machine. If the operator is not experienced or the machine cannot be trusted to regulate its own temperature, the thermometer is apt, about this time, to shoot up to 110 degrees and the whole incubator full of eggs to be destroyed. From this period to the end less artificial heat is required. In a warm room a large machine containing several hundred eggs will hold its heat for hours at a time without the application of any external heat whatever.

It is thought necessary to give eggs in incubators a daily airing, after the fashion of the hen. This is less essential when the hatching is done in a cold room. In airing eggs it is best to remove them from the machine in the trays and immediately close the doors so as not to lower the inside temperature.

While the eggs are being aired they should also be turned. Nearly all machines have devices for doing this, a trayful at a time, or automatically, by a clockwork contrivance, but in small machines it may be done by hand and the relative position of the eggs in the trays changed so as to better insure an equal chance for all. After the nineteenth day they should not be handled, except as the shells are chipped the broken side should be turned up.

PLATE IV.

DARK BRAHMAS.

CHAPTER V.

CARE OF YOUNG CHICKS WITH HENS.

Keep all chicks out of the wet grass in the early morning. It is not the wet feet, but the wet feathers that do the harm.

—Tim's Wife.

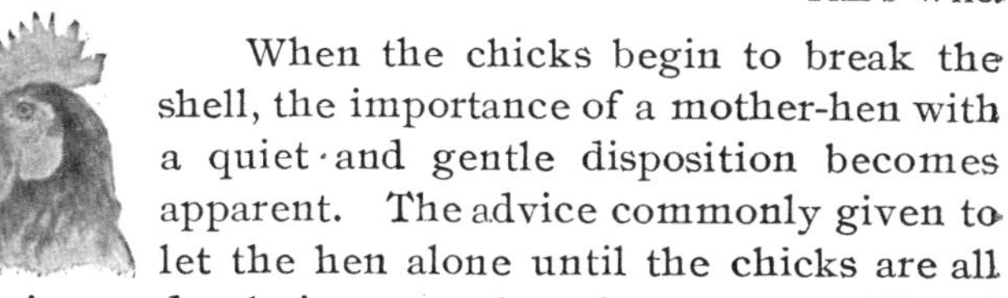

When the chicks begin to break the shell, the importance of a mother-hen with a quiet and gentle disposition becomes apparent. The advice commonly given to let the hen alone until the chicks are all out, is sound only in cases where hens are so wild and pugnacious that handling them will endanger the young, or the attendant is ignorant of the proper thing to do.

It is often good policy to take from the nest the chicks that come out first. This leaves more room for those that are to hatch, and when out of the nest they cannot be trampled on. This is especially wise when the mother is heavy, clumsy and fidgety and lacking motherly instinct. When several hens are hatching at the same date, it will often be found prudent, while the chicks are coming out, to transfer all the chicks and eggs from an unruly hen to those that exhibit more hen-sense.

All empty shells should be removed from the nest at once. Occasionally a chick is unable to get out after it has chipped the shell. The experienced hand can frequently give aid by carefully breaking the shell a little more, or tearing the tough surrounding membrane. Caution and experience are needed in the operation.

Eggs late in hatching are benefited by putting them for a few minutes in warm water tempered to about 103 degrees. If containing live chicks they will be seen to move in the water. If the chicks are dead they will remain perfectly still. After this warm bath the eggs should be put back at once under the hen without suffering them to become chilled.

Never in any case take all the chicks from the nest of a hen that is afterwards to be used as the mother of a brood; and if the chicks are of several colors, leave at least one of each color in the nest. Attention to these points will avoid trouble when the brood is returned to her.

Chicks taken from the nest should be put in a basket covered with woolen cloth, and placed near a stove. Do not remove from the nest until their down is dry. Such as show unusual weakness may be revived by pouring down their throats a few drops of warm, new milk.

Strong chicks need no food for twenty-four hours after hatching. If this time expires before it is convenient to return them to the hen, they may be fed in a box by a sunny window, and be put in their basket nest again until evening. The hen and her "sample lot" may, in the meanwhile, be fed near the nest. After dark the rest of the brood should be returned to her, and by the next morning mother and chicks are ready for the coop, which should be ready for the brood.

In cold weather it is best to set coops in an open

shed. They should always be set on a dry, slightly elevated location, so that they cannot be flooded by a sudden rainfall. Where the soil is at all wet they should be set on a platform made by nailing boards on two pieces of scantling. This platform should be of such a size that the sides of the coop will just fit over it. If allowed to extend outside of the walls the rain from the roof will keep the floor damp.

While the styles of coops are as numerous as their makers, the one here illustrated, having roof with double pitch and triangular ends, is as cheap and serviceable as any. To make it, take four pieces of 2 x 3 scantling, cut exactly 33 inches long and halved together at the top at such an angle as to make the base line of the front extend three feet. The coop is made two feet deep, thus giving a floor space of 2 x 3 feet. The roof may be covered by regular siding, or by fillistered barn boards cut into lengths of 2 feet 2 inches. The rear wall is boarded up solid, the front half way down, and the lower half is slatted. A loosely fitting door of boards may be hinged to the upper half to cover the slats and keep the brood in the coop when desirable. For summer weather, ventilation should be provided for by raising slightly the lower edges of the two uppermost roof boards, one on each side.

Here is shown a folding coop. The sides are hinged by iron pins seen at the dots on the upper front board in the cut. The solid rear end and slatted front are both hinged to the side and fold inward, which permits the sides to come together. When "knocked down" a coop occupies but little room when stored under shelter, as all coops should be when not in use.

Whatever the style of coop used, the chicks should be fed as soon as they are put into it. This is best done at first on a clean board laid on the floor or just in front of the coop.

As to what the first few meals should consist of, there is some difference of opinion even among practical poultry keepers. It is certain, however, that the traditional hard-boiled egg is not essential for the first, or for any other meal. When a hen steals her nest and brings off a brood, she feeds them successfully on weed seeds, insects and sundries until she brings them to the poultry yard and they can get the food fed to the rest of the flock.

Bread crumbs, moistened with sweet milk, are acceptable and nourishing for the first meal. Thousands are started every year on a mixture of corn meal and bran, half and half by bulk, scalded. It is well to scald this sometime in advance of feeding, and allow it to soak up the water and swell. It should be crumbly and not pasty. This mixture of corn meal and bran may be fed perfectly dry, and is so fed by successful poultry growers. A person of much experience uses bread crumbs and rolled oats, dry, the first week, and then for two weeks a mixture of equal parts by bulk of bran, middlings and corn meal, with a handful of meat-meal to the quart of the mixture. This is scalded an hour before feeding. If the bowels of the chicks are too costive he adds more bran, if too loose, more middlings.

Many make mixtures like the above into a stiff

batter with milk and baking powder, bake well and feed it dry. A woman who has been successful in this line gives her recipe for chick-bread as follows: take equal parts of sifted ground oats, corn and wheat, with wheat bran added equal to the whole bulk of ground feed, moisten with skimmed milk, add sufficient powder and bake. A little raw lean meat or finely cut raw bone and meat is beneficial. A little only should be given at first; a piece as big as a grain of corn is sufficient for a chick a few days old. This food is not essential when the grain ration is mixed with milk or dried meat.

In feeding chicks, as well as fowls, grass or vegetables should not be omitted. In the absence of grass in their runs, and in cold weather, chopped onions, lettuce, cabbage or other succulent vegetables should be supplied. Short clippings from the lawn, fresh, grassy sods, and the sweeping from the barn floor carried to their runs will be relished, and furnish the needed bulky vegetable food and afford healthful exercise. Little chicks should have five or six meals a day until three weeks old.

Gritty matter is required by chicks at the very beginning. To supply it, sprinkle coarse sand over the board on which they are first fed. If confined in houses or yards, or in runs where grit is scarce, it should be as carefully supplied as food. It is well to have a small trough or box in a convenient place filled with gravel, broken oyster or clam shells and granulated charcoal. The latter is not valuable as grit, but is very useful in correcting disorders of digestion.

WATER VESSELS.

Water should be given to the chicks from the start. It is best at all times to supply it in fountains from which they can drink but cannot get in with their feet. If supplied in open vessels they will foul it and contract colds, bowel disease or cramps.

FIG. 1.

A convenient water vessel for chicks may be made from an old fruit can and a flower-pot saucer, Figure 1. Cut a notch or punch a hole in the side next to the opened end, have the saucer just a little larger than the can, fill can with water, put on saucer and invert quickly. When chicks are older, the stone or earthen fountain shown here, Figure 2, holding a half-gallon or more, can be substituted. A very convenient fountain is shown in Figure 3, as the handle enables it to be carried around like a bucket. A tile fountain, preferred by some, is shown in Figure 4.

FIG. 2.

FIG. 3.

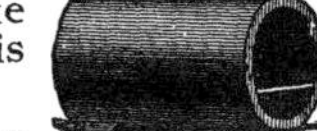

FIG. 4.

A common wooden bucket, cut down as shown in the cut, makes a first-class water vessel, convenient to carry. It should have a board over the top, or be placed under a stool to keep the water cool and to prevent the chickens from soiling it.

Before feeding ground oats and corn to little chicks sift out the oat hulls.

It is all right to have coops wind-tight, but all wrong to have them air-tight. Chicks must have ventilation as well as warmth. If insufficient air be admitted, the atmosphere of the coop becomes not only foul, but damp.

As soon as the brood is out of the coop in the morning, turn it up to the sun and air and spread dry earth over the floor. Whitewash the inside often. At midday turn down again. "Sweetness and light" applied to coops!

A strip of wire netting, one-inch mesh, two feet wide and about ten yards long, is "just splendid" for making a temporary yard for a hen and her young brood. Easy to put up, easy to move, and much better than the old style yard made of foot boards set on edge.

To make small runs for little chicks, make the sides of wide boards and cover with wire netting. This is better than making high fences. Old fowls cannot get into these covered runs and the chicks cannot crawl out through the wire, even if the mesh be wide.

PLATE V.

BUFF COCHINS.

CHAPTER VI.

CARE OF YOUNG CHICKS IN BROODERS.

Feed young poultry of all kinds early and late and often.
—Harriet.

LAUNCHED IN A COLD WORLD.

The rearing of chicks in brooders does not differ materially from the ordinary method, except that the intelligent instinct exercised by the hen in caring for her brood has to be exercised by the attendant.

Whether the chicks should be removed from the incubator soon after hatching or be left until nearly all are out of the shell, depends a good deal on the construction of the machine, especially of the egg-drawer. On this point the manufacturer should give explicit directions. As a rule, it is advisable to darken any windows that may admit light to the egg-drawer during the hatching process, to remove chicks as their down becomes dry, and all empty shells, but to open the incubator as little as possible. While the chicks are hatching the temperature is apt to rise but should not be allowed to go above 105 degrees. The removal of a basketful of chicks will cause the temperature to drop suddenly, a large amount of animal heat being thus withdrawn. Care must be taken to replace it by a surplus from the lamp. If the regulator at this stage fails to act, the chicks and eggs left in the machine may suffer a chill that will prove fatal. The attendant must, therefore, be very watchful at this time.

As soon as the chicks are dry there should be a brooder ready into which they may be put to remain for thirty-six hours, where they may learn to eat and run out and into the shelter of their silent mother.

The natural mother is just as warm when hovering her brood as when sitting on the eggs. The proper temperature of this first brooder must, therefore, be close to the hatching heat, say 90 to 96 degrees. This should be the heat of the center of brooder around which the chicks hover and from which they can move away when too warm. A brooder shaped like a box, that has warm corners, or that has a uniform temperature at all parts from which the chicks cannot escape is not safe. In a properly constructed brooder they quickly learn when too warm to move away from the heat just as they do from the body of the hen. They also learn where the source of heat is and will run to it when cold, but for the first two days it may be necessary to occasionally push them under cover to show them the way.

Instinct teaches the young bird to eat. The cluck of the mother hen and her pecking at the food calls attention to it and they follow her example. When feeding brooder chicks for the first time, it is only necessary to place them in the light and to drop the food before them in such a manner that their attention will be called to it.

For the first week the brood should be fed either in or beside the brooder and be confined near the heat so that they cannot stray away and become chilled.

Much of the sickness and mortality that befalls brooder chicks is due to chilling while they are very young, or from foul air and dampness in badly constructed brooders.

After ten days, the temperature of the brooder may be reduced to 80 or 85 degrees, and still lower in two weeks more. As chicks grow they generate more and more heat when they nestle together, and so require less in the brooder. When the weather becomes warm it may be necessary to shut off all heat in the day-time and during warm nights.

Manufacturers are prone to rate the capacity of their brooders too high. A brood of fifty is large enough no matter what the capacity of the brooder may be. Broods of one hundred can be handled until a month old, but after this stage is reached such a flock outgrows the largest single brooder or apartment. Much harm is done by the common practice of putting large numbers together.

Each brood of fifty chicks should have an outside run of not less than one hundred square feet in which to exercise until a month old. After this age they should have free range.

There are many kinds of brooders, some warmed by hot air, others by hot water; some furnish bottom heat, others top heat, and still others diffuse a current of warm air from the center outward. One of the latter is shown in Figure 1. Some are built for indoor and others for outdoor use; a double outdoor brooder is shown in Figure 2. In raising large numbers, single brooders in separate buildings are

FIG. 1.

used by some, while others prefer long houses containing many apartments, with an individual brooder in each. In these long houses some employ a greenhouse heating apparatus, warming the brooders by a system of hot-water pipes.

On general principles it may be said that bottom heat is practicable in mild weather only, when little artificial heat is required. Top heat, such as is obtained by radiation from a tank of hot water overhead, is unnatural and gives good results only when the tank is narrow and so placed as to prevent crowding into corners under it. The system nearest to nature is that which tempers the floor and the whole atmosphere of the brooder and gives off the greatest amount of warmth either by radiation, or by diffusing a current of warm pure air from the center.

FIG. 2.

It may be said in favor of long brooder houses containing many apartments that they are economical to build and manage; against them, that they are expensive to maintain unless run at full capacity.

In favor of individual brooders and small movable houses it may be said, they may be moved to new, clean ground whenever desirable; the flocks can be kept separate when disease comes to one part of the poultry yard; if fire breaks out in one house it need not destroy all, and when the birds are old enough the brooder can be removed, perches put in and the house affords a home for the flock until sold or moved to the hennery.

PLATE VI.

PARTRIDGE COCHINS.

Chapter VII.

GROWING EARLY BROILERS.

The early bird catches the worm.

Early eggs, early sitters ; early sitters, early chickens ; early chickens, early eggs and early profits.—Tim.

Broiler chickens are chickens of suitable size for broiling. The size established by convenience and custom is a weight of one to two pounds each. When much above this weight they pass as roasting chickens. Birds of this weight are tender and toothsome and are consumed mostly by persons who are able to pay well for the gratification of their tastes. The demand comes from wealthy private families and high-class hotels and restaurants.

The market for broilers opens soon after the New Year begins but is not at its best until asparagus appears. From the middle of March to the middle of June, a period of three months, there is generally a brisk demand for them. With the beginning of July, light-weight broilers are little called for, heavier weights are wanted, and as the weight goes up the price goes down, so that the poultry keeper finds it to his interest to keep his birds and feed them until they reach the "roasting" size, say six to eight pounds per pair. Growing broilers is winter work, as they must all be hatched and reared during the most unfavorable season for such operations. Hatching begins in November and ends with April, for the chickens, except such as are to be reserved for breeding, must all be in market soon after the last of June.

During the first half of this hatching period it is difficult to secure eggs of any kind, and especially such as are fertile and will produce strong chicks. The difficulty is the greatest just when the need for eggs is most imperative.

Provision must be made for overcoming this difficulty or the whole business will fail. To buy eggs in the general market is a very unsatisfactory method of obtaining them. They are apt to be stale or infertile, or from undesirable stock.

The only safe way to get good eggs is to own and feed the breeding stock, or to buy of those who know how to produce eggs for this purpose.

As the hens are to lay in winter they must be surrounded to some extent with summer conditions. This means that they must have comfortable houses, food suitable for producing eggs and plenty of exercise.

Whatever treatment hens may receive they will not lay well if moulting, nor if they have been put through a forcing process during the summer. The first eggs laid by pullets are of little value for hatching. The hens selected for making up the breeding stock should be well over their moult, not too fat and in good health. If pullets are chosen they should be from the early broods.

The hens most likely to meet the requirements of the case during November and December will be found among those hatched late in the previous summer and fall. By the time these are exhausted the older hens and early pullets will be ready to continue the egg supply.

Suitable hens having been secured they should be mated with early-hatched cockerels.

Since the work of caring for the chicks in winter weather is arduous, and as prices decline rapidly after a certain date, it is of much importance to the poultry keeper to have chicks that grow to the proper size in the least possible time. There is a difference in breeds and crosses in respect to quickness of growth. Some will attain to a merchantable weight in eight weeks, while others will require from ten to sixteen weeks.

Among the pure breeds that make quick-growing broiler chicks may be mentioned Wyandottes, Plymouth Rocks and Light Brahmas. Leghorns grow quickly to the broiler stage, but are rather small. They make a good cross with Brahmas and Cochins, Leghorn males being mated with the Asiatic hens.

As broilers when they are dressed for market are in the pin-feather stage, it is desirable that these feathers should be light in color, for if dark the smallest one left on the carcass is apparent, and the large ones when plucked leave a stain on the skin. For rearing broilers, therefore, fowls of light plumage, other qualities being equal, should always be chosen. Buff-colored fowls have light colored pin-feathers and are always safe to use for this purpose.

When the appearance of the carcass is not a matter of importance it is safe to use any Mediterranean-Asiatic cross. Houdan males may also be used with Asiatics or with Dorkings with good results. A Plymouth Rock or Houdan cross with a breed having any black in the plumage is apt to produce progeny with solid black plumage. To secure both light pin-feathers and the yellow skin so much prized in some markets, a White Leghorn-Buff Cochin cross will fill the bill. A White Plymouth Rock-Buff Cochin cross is also to be commended both for broilers and larger roasting chickens.

The hatching of broiler chicks on a large scale must be done with incubators, since but few hens are broody in fall and early winter. The brooding must also be done in artificial mothers, and for the most part, under cover of a good roof.

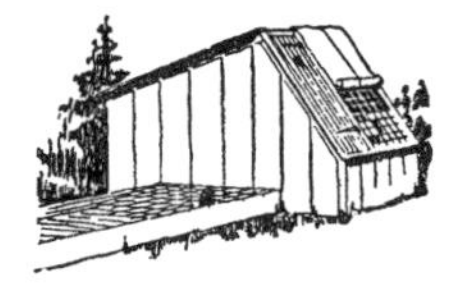

An individual brooder house in common use among broiler raisers is shown here. It is five feet four inches by eight feet on the ground. The roof in front is divided into two parts, three feet are covered by wire netting and over this cotton cloth which may be rolled up when weather permits; the other part is the door for the attendant. The rear wall is three feet six inches and the front one foot nine inches. A yard four by sixteen feet extends from one side.

A section of a good type of a long house is shown in perspective at Figure 1. It is eighteen feet wide, divided into pens three feet wide, each containing a brooder designed to hover fifty chicks. By reference to Figure 2 it will be seen that the glass run is shut off from the house by a solid, hanging door that swings inward against the front wall. This is opened in the

FIG. 1.

day-time to give plenty of sunlight, and closed at night to shut out the cold that enters through the glass. It is seven feet high in front and four and one-half at the back. The passageway is at the rear and is sunk sixteen inches, thus allowing the building to be made low without compelling the attendant to stoop. The brooders are set along this passageway in such a manner as to bring their floors on a level with the floor of the house. Light and ventilation are both supplied from the rear as well as from the front wall.

FIG. 2.

One who has raised thousands of broilers successfully gives his method of feeding as follows: "I give no feed for thirty-six hours, and don't allow them to go more than a foot from the brooder.

"For the first two weeks I feed them cake made as follows: two quarts coarse corn meal, one quart bran, one quart middlings, one teacup ground meat (be sure that there is no pork or fish about it), one cup fine bone, wet with a scant pint of water. The secret in making this cake is in not getting too much water in it and in baking it thoroughly in a quick oven. Feed three times a day all they will eat up clean in a short time. Overfeeding is a cause of bowel trouble. Give them all the water they want, with the chill taken off.

"After they are two weeks old I take one quart of corn and oats sifted, one quart bran, one pint each of middlings and coarse corn meal, a cup each of meat and bone, moisten with hot water and let it stand a short time. I add some of this to the cake gradually until they are three weeks old, then I drop the cake and feed the other until they are six weeks or two months old. Then I take two quarts corn and oats ground, one quart corn meal, one quart middlings, one pint of bran, one pint each of bone and meat, wet with hot water, using more water than for the small chicks. Let it swell before feeding.

"Charcoal is very necessary to keep chicks healthy. Have it ground fine and keep before them all the time, also ground flint. I hash them up onions and cabbage occasionally.

"Don't let the chicks run out in the yard in winter until they are a month old."

SPRING CHICKENS.

Rub off the dusty windows and let in the light.

Lettuce affords a quick-growing and choice green food.

The market has never yet been overstocked with broilers.

A thrifty chick will weigh one pound when six weeks old.

It does not pay to feed runts. Weed them out and fertilize the garden.

Dry earth is the best and cheapest disinfectant and deodorizer obtainable. Store plenty of it.

If you can't get milk and can get creamery whey, use it. While not equal to milk it is a good substitute.

Raw chopped onions fed at night are said to be a safeguard against roup. They are wholesome at any rate.

Let the flock have a space on the ground somewhere covered with litter, and keep them in a state of activity.

Pour boiling water on wheat and let it soak over night. Give the broilers in the fattening coop an occasional feed of it.

Cash in the pocket is not in danger of gapes, cats, crows, rats, roup or cholera, and therefore is better than the chickens in the coop, if they are old enough for market.

Boiled rice, sweetened with brown sugar, is excellent for putting the finishing touches on the early broilers. Give them one or two meals a day for a week before sending them to market. Broken rice can be bought cheap.

PLATE VII.

LANGSHANS.

Chapter VIII.

HENS EXPRESSLY FOR EGG PRODUCTION.

The best " egg producer " is good food and plenty of it.
The hen that sits on the roost or fence in zero weather, or stands on one leg in the snow all day, is not a winter layer.

—Harriet.

Keeping hens for laying purposes chiefly is a profitable part of the poultry business when rightly conducted and when the surrounding conditions are favorable.

The selection of the laying stock is a matter of much importance. There is the "laying type" among hens just as there is the " milk type " among dairy cows. These are found to some extent among all breeds but in larger proportion among the Mediterranean class. Generally speaking, good layers are fine-boned. This is seen in the shank which is slender and relatively short. This feature is determined by comparing specimens of each breed by themselves, that is, Leghorn hens must be compared with Leghorn hens; Brahma hens with Brahma hens, etc. A small feminine head with prominent eyes and a slender neck are also indications of a good layer, just as similar features in a cow betoken a copious milker. The body of a good layer is rather long and wedge-shaped, smaller in front than back.

ONE OF THE "LAYING TYPE."

The good layer is of a lively, active, restless disposition, ready to play or fight with her companions and always in search of something to do or to eat.

Any one who has been a careful observer of hens will recognize the business hen as soon as his eyes rest upon her. Hens of the opposite character are just as readily detected by their coarse-boned shanks, thick necks, masculine heads and masculine make-up.

The breed to be chosen for layers will depend partly on the taste of the poultry keeper, to some extent on the market in which the eggs are to be sold, and on whether the owner wishes to combine meat production and the sale of pure-bred eggs for hatching

HOUSES AND YARDS OF A FANCY POULTRY RAISER.

with the market-egg business. The breed that everybody pronounces "best" for laying or for any other purpose has not yet been discovered. Some prefer pure-bred hens, others crosses.

It is generally conceded that the Mediterranean breeds lay the largest number of eggs. Their eggs are mostly white or but slightly tinted, and have thin shells. Their color is objectionable in some markets and their fragile shells render them more liable to break in shipping. When the surplus hens have to be marketed for meat, they do not make first-class dressed poultry. What, therefore, is gained in the number of eggs may be partly or wholly lost in selling the dressed meat. This is the argument on one side. On the other side it is maintained that the small breeds seldom become broody, mature quickly and come quickly into profit, and that these facts combined with the increased number of eggs laid, compensate for any loss in weight or price of carcass.

Those who combine the raising of broiler and roasting chickens or capons with the production of eggs, generally choose the American breeds or crosses.

When a poultry keeper can find sale for pure-bred eggs and fowls in connection with his egg business, the breed that is most popular with buyers is the breed he is apt to prefer.

Those who have a special or private trade for darkly tinted eggs should select Wyandottes, Plymouth Rocks, Rhode Island Reds and Brahmas.

Cochins, Poland and the English class are seldom chosen for stocking an egg farm.

Whether a poultry keeper shall raise his own hens or buy them, depends on various circumstances. Fully one half of all chickens raised will be cockerels. If, therefore, five hundred pullets are wanted, one thousand chickens must be raised, and more than this must be hatched, for some will always die before reaching a marketable size. Some who practice the rearing of their own layers, give at least plausible figures to prove that the sale of cockerels yields enough profit to pay for the raising of the pullets to the laying age, so that they cost practically nothing. This may be a rosy view, true only in certain favorable conditions. It is undoubtedly true that those who grow their own stock can have the kind they want, and are not compelled to take a motley collection such as can be gathered by promiscuous purchase.

The managers of some of the large egg farms furnish eggs for hatching, from such stock as they choose, to farmers in the surrounding country to hatch for them, and buy the pullets at a certain age and price agreed upon between the contracting parties. This plan works well, as it leaves the operator free to give his entire time to the care of the layers, and also permits him to conduct his business on a smaller area and with less capital. For without the rearing attachment less land, fewer buildings, and less labor are required.

The most successful hen farms consist mainly of houses with yards of only moderate size. Free range is not a necessity for hens kept chiefly for eggs. It is stoutly affirmed by those who have had experience with both methods that with proper care a flock will produce a fifth more eggs in confinement than when at liberty. Greater care is required with shut-in hens, but there are compensating advantages: they are under the attendant's eye at all times, are easily controlled, fed and tended, and out of danger from enemies, and cannot commit depredations on the field or garden crops of their owner or his neighbors.

Except in sections where land is low in price and deep snow does not fall, the plan of colonizing hens in small houses scattered over many acres, and giving them free range, is not at all feasible when the object is to produce market eggs.

The style of house most economical to build, and that best serves its purpose on an egg farm, is a long, low shed-roofed structure, divided into apartments and facing south or southeast. Several typical buildings of this description are owned by a noted egg farmer. They are each two hundred and sixteen feet long, ten feet wide, seven feet high in front and four and one-half feet in the rear. The front leans back one foot, making it exactly ten feet wide, saving two hundred and sixteen feet of roofing, and giving the windows a slant so as to get a stronger sunlight on the floor. Hemlock frame and boards are used, the front battened and the roof and rear wall covered with tarred felt. The interior is partitioned off every twenty-four feet, giving two hundred and forty square feet of floor space to each apartment. There are two large windows to the front of each room, these are made to slide and serve also as doors into the yards in front. The partitions are boarded up three feet, and wire netting used above the boards. There is a gate two feet wide on the front side of each partition, hung with double-acting spring hinges, so the attendant can walk right through with two pails of feed or water without stopping to open or close them. A platform twenty-eight inches wide, two feet above the floor, runs along the rear of each room, and ten inches above this platform is a perch. The nests are placed underneath the platform on the floor, the hens entering from the rear. These houses all have earth floors.

Each apartment accommodates thirty to forty hens, and each flock has a yard in front of its apartment twenty-four by sixty-four feet, in which are growing one or two peach or plum trees. These houses for convenience, cheapness and practical business cannot easily be excelled.

The general rules of feeding given in Chapter II when treating of the best method of getting fertile eggs for hatching, will apply in this case.

It will, however, be entirely safe in feeding hens for market eggs alone to force them a little harder by feeding more highly seasoned and more nitrogenous foods than would be advisable when hatching eggs are wanted.

On every egg farm there should be a large boiler or steam cooker for cooking vegetables and making compounds of meat, ground grain and vegetables. A good morning ration may be made of equal parts of corn meal, fine middlings, bran, ground oats and ground meat. This should be stirred into a pot of cooked vegetables while boiling hot until the mass is as stiff as can be manipulated by a pair of strong arms. Seasoned with salt and cayenne pepper. Potatoes, beets, carrots, turnips, onions or any vegetable clean and free from decay will be acceptable. Cut clover hay may be substituted for vegetables for an occasional meal. The above contains a variety of food elements such as compose the egg, bone and muscle of the hen, the fat-forming elements not being prominent. For the noon meal, wheat is the best single grain. It may be mixed with good oats and scattered in chaff or leaves on the feeding floor. The night feed should be a light one, consisting of whole corn.

Plenty of fresh, clean water is just as essential as food. Sharp gravel or grit of some kind is as much needed as food and water, and should be accessible at all times.

Green bones and meat shaved in the modern bone cutters is a prime article for laying hens. It may be fed to advantage in place of the ground, dry meat, three days in the week, an ounce to each hen. Those who are near large cities can sometimes get cooked lean meat and bone from bone-boiling establishments. This is an excellent form of meat for use in cool weather. All forms of meat should be fed cautiously, a little at first and more as the fowls become accustomed to it.

Some of the most successful persons in this business have land in addition to their poultry yards and raise a considerable portion of the food the hens eat.

The farm is run in the interests of the hens. If cows are kept the skim-milk is fed to the hens. All vegetables except such as are used in the family, or are extra fine and command an extra price, find their way to the poultry yard. Clover, oats, wheat, rye and corn fodder are harvested green, run through a fodder cutter and fed to the hens. Cabbages are raised and buried, turnips and beets are grown and stored for winter feeding. Any clover not needed for summer feeding is cured and used in winter.

For the health of the hens, and to secure the largest egg production, it is necessary to furnish an abundance of succulent and bulky food along with the more concentrated grains and meat. It is cheaper to raise this than to buy it, while the grains and meat can probably be bought cheaper than they can be raised.

On an egg farm the most exacting labor is required in winter, for the wise manager aims to produce winter eggs, since prices are then at their best. Summer is a season of comparative leisure in the hennery and the extra help required in the winter may be profitably employed on the farm in growing necessary supplies.

As to how long hens should be kept for laying authorities do not agree, but it is doubtful if they should ever be retained long after they have passed the spring months of their second year. During spring and early summer dressed hens command good prices. So fast, therefore, as they show signs or breaking down with too much fat, quit laying and become broody, they should be started on their way to market. By midsummer the stock in the houses should be reduced to one-half or less of the full winter complement and consist only of the best of the yearlings.

Cut green clover fine, and feed it to all fowls confined in yards. Splendid.

Observe how a flock will nestle on a well-littered floor in winter. A hint to the wise.

Snow is a poor substitute for water. Fowls should not be compelled to eat it to quench their thirst.

This is a fairly well balanced daily ration for one hundred hens:

Clover Hay	2.74 lbs.
Potatoes	2.74 "
Corn Meal	5.48 "
Ground Oats	2.74 "
Cottonseed Meal	.274 "
Barley Meal	2.192 "
Total	16.166 lbs

or, in round numbers, sixteen pounds of the mixture.

PLATE VIII.

SINGLE COMB BROWN LEGHORNS.

Chapter IX.

THE FARMER'S FLOCK.

Give the hen a good chance to scratch and she will raise that mortgage for you.

A hen will eat anything a hog will eat and make a good deal better use of it.—Tim's Wife.

The larger part of all the eggs and poultry sold in the markets of the great cities and smaller towns comes from the farmer's flock. The amount from each is small, but the aggregate immense. When proper attention is given to this flock the profit is as large, if not larger, than from any other part of the farm operations.

The mistake of keeping too small a number of fowls is sometimes, though rarely, made. Fowls, with their omnivorous and voracious appetites, are excellent scavengers, and if allowed the privileges of the premises will utilize much that would otherwise go to waste. This wastage on large farms is sufficient to supply a flock of one hundred laying hens three-fourths of all the food they need; if but ten or twenty be kept there will be more or less loss.

The much more frequent mistake is made of overstocking. The wastage is consumed, the crops in the vicinity of the buildings are destroyed, large quantities of grain in addition are fed, the houses are crowded to suffocation, and the ground in the entire circle of the farm buildings becomes befouled. All may go well for a few years and then disease invades and disaster comes, and the farmer arrives at the conclusion that there is no profit in chickens.

The size of the flock should be regulated by the circumstances surrounding each case. Large stock farms where large quantities of grain are used, where there is plenty of grass, numerous shelter-sheds and no truck gardens near the house, furnish favorable conditions for keeping a large flock with profit. Dairy farms, also, where grass and skim-milk are available, will support economically a big flock of laying hens or grow capons of good quality. One who follows trucking or small fruit growing must limit his flocks or confine them in yards during the growing and fruiting season, which adds to the expense and care. If properly managed the expense and care will be repaid, because on such farms there is a considerable offal that can be utilized for poultry food.

PETER TUMBLEDOWN'S POULTRY HOUSE.

Too little care is given by the average farmer to the breeding of his flock. The quickest way to raise the standard of such a flock with little expense is to cull out and sell old, broken down, scrubby and inferior specimens and mate the balance with pure-bred males. If it is desired to increase the size use males a *little* larger than the common stock. Very large males should never be used with small or medium hens. If the hens are large and heavy use a male a little smaller. This process may be continued to advantage each year, but always use pure-bred and never the cross-bred males. The pure-bred birds may be hatched from eggs bought or they may be purchased late in summer or autumn from breeders who will sell such as are slightly off color, or have some slight defect in comb or in other minor points that do not affect their value as a farmer's fowl.

In planning and erecting farm buildings too little attention is given to providing proper shelter for poultry. While elaborate and costly structures are not required, they should be storm proof, free from drafts in cold weather, have ample ground-floor space, and be convenient for the attendant.

FIG. 1.

The last point should not be overlooked, since a very little saving of time and labor each day of the three hundred and sixty-five, amounts to a considerable saving in the year, and this may be accomplished by a small additional outlay at the start.

The style, size and cost must be determined by the builder's needs, taste and pocket-book. There is no "best" house for all situations and all persons. A few are given rather as suggestions than as models to copy.

The style illustrated by Figure 1 is economical of

lumber, as it consists chiefly of roof. It will be an advantage, especially on low ground or clayey soil, to have the floor filled in six or eight inches deep with cinders or broken stone and covered with gravel or sand. The ventilator is for summer use alone and should be tightly closed in winter. The cut represents a house twelve by sixteen feet, set on a wall two feet high, the point of the roof being eight feet above the floor.

Figure 2 exhibits a good type of house for general use. As will appear from the illustration it has two enclosed apartments with an open shed in the center. Both the apartments being raised thirty inches from the ground the whole floor space is available as a scratching-room. The house is twelve by twenty-four feet, the shed and end parts being eight by twelve feet each. One end is the roosting-room, and the other the laying and hatching-room. The fowls reach these rooms when the doors are shut by means of cleated boards extending from the ground to an opening in the floor. A passageway from one to the other eighteen inches wide and enclosed by wire netting is shown in the cut along the rear wall of the shed. Figure 3 shows the plan of this house.

FIG. 2.

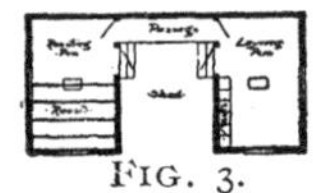

FIG. 3.

A serviceable and good all-around house is shown at Figure 4. A good width for a building of this character is eighteen feet, this allowing three feet for the hall or passageway in the rear, nine feet for the main house and six feet for the scratching-room or shed.

FIG. 4.

Figure 5 shows how this house may be divided for two flocks. The nests are accessible from the hall. It is always convenient to have a yard of generous dimensions, securely enclosed, in which it is possible to confine the flock while crops are young, or whenever desirable to do so. This yard should be large enough to plow with a horse and be planted with plum or peach trees, and grape vines to afford shade in hot weather, and for growing fruit.

FIG. 5.

The matter of fencing the poultry yard may be left in the hands of the owner, with the suggestion that it is cheapest in the end to build a substantial fence at the start. The cheapest temporary and movable fence that can be erected is one of wire netting. This should have posts every eight feet, a board at the bottom, but no rail or board at the top. The posts need not be heavy.

The farmer's flock should have as careful feeding and attention as any other stock on the farm. To insure such attention some one member of the family should take the matter in hand and make it his or her business. Regularity in feeding is essential to the best results. Economical feeding means that all the wastes of the family table, the dairy, the garden and the field should be turned into eggs and poultry meat.

NEST NOTES.

A HANGING NEST.

Use small hens to hatch thin-shelled eggs.

The best feed for sitting hens is corn. They should have clean water and gravel and access to dry earth. They need little else.

If the hen deserts the nest for a few hours and allows the eggs to become chilled, do not throw the eggs away. Let them have another trial; they will stand exposure for a long while and yet hatch well.

Whatever else you do, don't follow the stereotyped advice in poultry books and papers to make the nests of sitting hens on the ground--not, at least, before June.

In April showers look after the young broods. A "saturated solution" of chicken is N. G., except for soup.

Boil beef or pork cracklings with small potatoes, add corn meal, mash all together and make a dish fit for the chickens of a king.

The most acceptable lays of spring are furnished by the hens.

It is bad policy to keep the big, slow-motioned fowls and the small, nervous, quick-motioned breeds together in one flock. They require different feeding and treatment; they do not harmonize.

A hen's teeth are in her gizzard. Sand, gravel and like substances are the teeth. Keep them sharp.

A state of fear and excitement is unfavorable to egg production. Every movement among a flock of hens should be gentle.

The wide-awake poultry keeper is up and around among his flocks early in the morning and late in the evening.

Drinking water in cold weather should be neither hot nor ice-cold, but simply cool, and always clear and fresh.

A GENERAL-PURPOSE HEN.

PLATE IX.

SILVER POLISH GOLDEN PENCILED HAMBURGS

CHAPTER X.

THE VILLAGE HENNERY.

In cold weather keep your eyes open and the cracks in the hen house closed.—Harriet.

The hen turns grass into greenbacks, grain into gold, and even coins silver out of sand.

Persons living in towns and villages may oftentimes find pleasure and profit in keeping a small flock of poultry. The mistake most frequently made by those who undertake to do so is in attempting to keep too many. When confined in small yards they become unhealthy and unproductive; if permitted to roam they become a nuisance in the neighborhood and a prolific source of unneighborly feeling and of disputes which only a justice of the peace can settle.

To maintain a peaceful mind and a quiet community attention should be paid to the variety of fowls kept, and to the yard fences. The Asiatic breeds are particularly fitted by their quiet nature and indisposition to rove for stocking a village hennery. They not only thrive better in close confinement than the smaller and more active breeds, but are more easily confined. A fence four feet high will restrain them. If the fence be made of wire netting, a six-inch fence slat at the bottom and three feet of netting above it will be sufficient. Temporary runs can be made for them in the garden or anywhere, by driving down stakes and attaching yard-wide netting.

The tendency of these fowls is to lay on too much fat, but this can be regulated by feeding but little grain or other fattening foods, and compelling them to exercise by scratching in leaves, chaff or soil. The bane of small flocks is overfeeding and this must be avoided to get the best results from Asiatics. All scraps from the table that commonly go to prowling dogs and cats should be fed to the chickens. Milk or other liquid wastes may be mixed with bran. They should have a liberal supply of grass from the lawn, and waste green vegetables from the garden, and only a small ration of grain. Thus fed they will lay and will not grow fat. Lawn clippings, dried in the shade and stored in bags make the choicest of winter greens for a village flock, or indeed for any fowls.

Those who prefer the smaller and more active breeds must provide higher fences, or where the runs are small, make the fences low and cover the entire top with netting. Sometimes in towns where the houses are crowded close together it is difficult to get sufficient sunlight and air in the poultry yard to render the quarters dry and healthful. In such cases high picket fences make the difficulty still worse, and wire netting is much to be preferred both for utility and appearance.

While an expensive house is not a necessity in a town, it need not be rude and unsightly. Some simple ornamentation is within the reach of nearly every one, and if it could be more generally applied would greatly add to the attractiveness of a rear view of a village street. The house on page 76 may be called the *Farm Journal* village poultry house. It represents a structure ten by twenty feet, with eight feet of the length enclosed and twelve feet left open for a shed. The interior arrangements as well as the size and exterior ornamentation may be left to the needs and fancy of the owner.

FIG. 1.

Another building well adapted for a small flock is shown by Figure 1. This house is ten by fifteen feet, five feet high in the rear and seven feet in front, with a hood or overshoot. The roosting-room occupies five feet of the length and is elevated two feet from the floor. A board along the front keeps in any litter that may be thrown into the shed. Such a house permits the flock to live out of doors and to enjoy plenty of air at all times. During stormy weather they may be confined to the house by covering the front with a screen of wire netting. The plan of this house is shown in Figure 2.

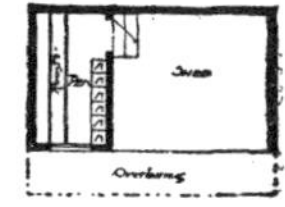

FIG. 2.

A flock of Bantams will be found useful where room is limited. Although their eggs are small, they are prolific layers. The birds themselves being small do little injury to lawns or gardens when at liberty, while they destroy many harmful insects.

BANTAM HOUSE.

The small, portable house and run here illustrated is admirably adapted to accommodate a flock of these little beauties. The netting door is divided so that the top of it may be opened by the attendant and feed and water put in the run, without entering or letting the chicks out. The whole structure should be made of light material and of a size to render it easily movable by two persons of ordinary strength.

As the purpose of keeping poultry by the average villager is to supply his own table with eggs and poultry but few chicks should be hatched. These should be kept separated as much as possible from the flock of fowls; colonized, if possible, in a different quarter until ready for the table or to take the place of the laying stock in the common runs.

When no hatching eggs are required no males should be kept in the flock. They are useless boarders and will soon "eat their heads off," and should themselves first be eaten.

The rooster, speaking botanically, is the crow-cuss of the poultry yard.

Dump old mortar and broken plaster in the poultry yard.

Damaged grain may be used if scorched slightly before feeding.

Puny, sickly birds are only profitable for fertilizing trees and vines.

Pull out the feathers in one wing to prevent flying.

A good cat and vermin-proof coop for the village hennery is often necessary. A simple one is shown herewith.

As an egg persuader, try equal parts of bran, corn meal and ground oats, mixed with one-eighth part of linseed meal; that is, four quarts of the linseed meal to one bushel of the grain mixture.

If snow that falls on the roof is likely to melt and drip through, shovel it off. A shower bath of snow-water means roup and death later on.

PLATE X

HOUDANS.

Chapter XI.

BREEDS OF CHICKENS.

Don't estimate the size of the egg from the length of the cackle
Fine birds are not made by fine feathers.—Harriet.

The American Poultry Association publish a book called the "American Standard of Perfection," that contains a description of all kinds of fowls the Association thinks worthy of recognition as pure-bred poultry. The descriptions of this book are of ideally perfect specimens of the kind, and are intended to set up a model for breeders to follow.

WHITE PLYMOUTH ROCKS.

A scale of points is given by which fowls are compared with one another and by which they are judged and rated when on exhibition. Certain disqualifying marks are also mentioned that exclude all specimens having them from competing at exhibitions.

This "Standard of Perfection" is, in fact, the

fancier's text-book. Thoroughbred chickens are divided into ten general classes.

The first is the American class. This includes five breeds: Plymouth Rock, of which there are four varieties — the Barred, Buff, Pea-Comb, and White; Wyandottes, of which there are five varieties —Silver-Laced, Golden, White, Buff and Black; Javas, in three varieties—Black, Mottled and White; American Dominiques and Rhode Island Reds.

Of this class, Plymouth Rocks and Wyandottes are the most numerous and widely known.

The characteristic shape and appearance of the Single-Comb Barred Plymouth Rock is well exhibited in colored Plate I. The color of the plumage is a grayish-white, each feather crossed with bars of blue-black. The color is the same as that of the Dominique. They are a general-purpose chicken, are superior layers and make shapely dressed poultry. Being well adapted to farm conditions they have long been popular as the "farmer's fowl." A full-grown cock should weigh nine and one-half pounds, and a hen two pounds less. The other varieties of the breed differ only in comb or plumage.

WHITE WYANDOTTES.

Wyandottes, a breed of more recent origin, have

also a reputation for general usefulness. The Silver-Laced are shown in colored Plate II. As will be seen they are compactly built and make a fine appearance as dressed poultry, at whatever age they may be killed for market. A mature male should weigh eight and one-half pounds, and a hen two pounds less.

Javas and Rhode Island Reds have some peculiarities of their own, but are similar in size and other respects to Plymouth Rocks.

BLACK JAVA PULLET.

Dominiques have rose combs, a neat, trim shape and a gray, hawk-colored plumage. In size they rank with the Wyandottes. They are the oldest American breed and it was from a cross of these, with a larger breed, that the Plymouth Rocks originated.

The second general division is the Asiatic class, which includes Brahmas, Light and Dark; Cochins—Buff, Partridge, White and Black; Langshans—Black and White.

Light Brahmas, illustrated in colored Plate III, are the largest of all the breeds. They are a modification, by careful breeding for many years, of the old Brahma Pootras. As now bred they are a noble and attractive fowl and have also great practical merit. As layers they equal, if they do not surpass, any large fowls. For making heavy broilers at eight and ten weeks of age they are among the very best. After they are three months old they do not make first-class dressed poultry until well matured, on account of their rapid growth and bony frame. The standard weights for matured birds are twelve pounds for cocks and nine and one-half pounds for hens, but they frequently exceed these figures.

Dark Brahmas are shown in colored Plate IV. These are usually a pound lighter in weight than the Light Brahmas, and while they have the Brahma carriage, their shape resembles their Cochin cousins, thus betraying their probable origin in a Light Brahma-Cochin cross. There is a marked difference in the plumage of the male and female. When carefully bred to feather a flock of Dark Brahmas presents a very attractive appearance. They have for many years been highly prized for market purposes, especially by those who grow capons.

Buff Cochins, colored Plate V, are the old yellow Shanghais with their stilted legs and long necks reduced by careful breeding. The illustration is a faithful likeness of well-bred Buffs of the present-day type. They have no more neck or length of leg than seems absolutely necessary, their bodies are blocky and covered with an abundance of soft, fluffy plumage of a creamy, golden hue. Their plump form and yellow skin make them popular with market poultrymen. In disposition they are gentle, quiet, even lazy, and are easily restrained. They are only fairly good layers, but are persistent sitters and good mothers. Since their introduction into this country Buff Cochins have probably been used for crossing upon the common stock of farmers to a greater extent than any other single breed. The standard weight of mature birds of the breed is, for cocks, eleven pounds; and for hens, eight and one-half pounds.

The other varieties of Cochins differ only in color. The Partridge Cochins are admirably represented by colored Plate VI, a reproduction from life of superior specimens of the variety. The plumage is very beautiful, being like that of the famous Black-Breasted Red Game, and suggests an origin in a cross of Game and Cochin.

Langshans are the latest accession to the Asiatic class, having reached us by way of England. As will be seen by referring to colored Plate VII, they have a shape and carriage peculiar to themselves. Their plumage is abundant but not so fluffy as that of the Brahmas and Cochins. The plumage of the Blacks is a glossy black, showing a beautiful greenish metallic sheen when viewed in a good light. Langshans are considered to be the best layers of their class; although their skin is white they are a good market fowl and their meat of superior quality.

The third class is the Mediterranean. This embraces Leghorns of which there are eight varieties—Brown, Rose-Comb Brown, White, Rose-CombWhite, Black, Dominique, Buff, and Silver Duck-Wing; Minorcas—Black and White; Andalusians and White-Faced Black Spanish.

BUFF LEGHORNS.

Of these, the Leghorns are the most widely disseminated and most numerous. The Single-

Comb Browns are well illustrated by colored Plate VIII, which exhibits, also, the general type of the breed in respect to shape and carriage. They are smaller than any of the American class, sprightly, active, light of wing, early to mature and famous for laying the greatest number of eggs of any of our domestic fowls. Their eggs are of medium size, but large in comparison with the hens that lay them.

The brooding propensity has been bred out of the whole class to a great extent, and they are commonly referred to as non-sitters. This is only relatively true, for the best-bred hens among them will occasionally become broody. It is, however, true of all that they cannot be depended on for hatching and rearing chicks.

BLACK MINORCAS.

The Minorcas have a general resemblance to Leghorns, but have longer, deeper and heavier bodies. The weight of a full-grown male should be eight pounds, and that of a female six and one-half pounds, which is fully a pound heavier than Leghorns commonly reach.

Minorca hens are famous for producing large numbers of eggs, and when they have attained the age of two years and over the size of their eggs is quite remarkable.

The White-Faced Black Spanish are a distinguished looking fowl, and may appropriately be classed with the Light Brahma as belonging to the aristocracy of the poultry yard. While having the general characteristics of the class, their white face, black, silky-glossed plumage, a body of peculiar shape set well up on long, slender legs gives them an appearance quite distinct from all others.

WHITE-FACED BLACK SPANISH.

They lay a large, creamy white egg. Andalusians might be called Blue Leghorns. They are a beautiful fowl, but for some reason are not largely bred.

A fourth class is the Polish, which embraces eight varieties, namely—White-Crested Black, Golden, Silver, White, Bearded Golden, Bearded Silver, Bearded White and Buff-Laced. The Silver Polish, and the general appearance of the breed, are seen in colored Plate IX. Both fowls and eggs of this breed are rather small and are mostly bred for fancy purposes. They are prolific producers of rather small eggs, and very pretty.

WHITE-CRESTED BLACK POLISH.

The fifth class is the Hamburg. This includes the Hamburg breed with six varieties—Golden-Spangled, Silver-Spangled, Golden-Penciled, Silver-Penciled, White and Black; Red Caps; Campines—Silver and Golden. The Hamburgs are the principal breed and the Golden-Penciled are shown in colored Plate IX. Hamburgs, like the Leghorns, are celebrated as egg producers, but their eggs are small, like the fowls. They have been used with good effect to cross with larger fowls, to increase their laying quality.

SILVER-SPANGLED HAMBURG.

The Red Caps are a larger type of Hamburg with a very large rose comb. The Campines, a variety of recent introduction, are similar in general appearance to Hamburgs, but have single combs.

The sixth class embraces the French breeds: Houdans, Crevecoeurs and La Fleche. The Houdans are shown in colored Plate X. They are distinguished by a large crest, V-shaped combs and plumage of mottled black and white, the black predominating. A full-grown male should weigh seven pounds, and a female six pounds. Houdans are good layers, have compact, well-proportioned bodies, and are superior table and market fowls. The flesh of all the French breeds is white, the bones are small and the meat juicy. Like the Dorkings, they have five toes on each foot. The Crevecoeurs and La Fleche have black plumage and are larger than the Houdans. For some reason they have not become popular in this country and are not so well known as the latter.

The seventh class is the English breed, the Dorkings, of which there are three varieties—White, Silver-Gray and Colored. Colored Plate XI is a good representation of the Silver-Grays. The Dorkings have a characteristic shape that is well shown in the cut. The body is long, deep and full, neck and legs short, and the whole appearance solid and substantial. The standard weight of mature males of the Silver-Gray variety is eight pounds, and of mature females six and one-half pounds. Colored Dorkings should weigh a pound heavier. These all have white flesh. They are good layers, but are especially prized for their market and table qualities. The Orpingtons, several varieties classed according to color of plumage, have of late years been imported from England. They are good layers and good table fowls, and in size compare favorably with our Plymouth Rocks.

The eighth class comprises—Games, Game Bantams, Cornish Indian Games and Malays. There are eight varieties of Games and a corresponding number of Game Bantams. The typical Game shape is well exhibited in the Black-Breasted Red Game Bantam in colored Plate XIII. They all have single, erect combs and wattles, but it is the fashion to cut these appendages off. It is this operation, called "dubbing," that produces their fierce and war-like appearance. Contrary to a common impression the varieties of Games named in the "Standard" are seldom ever bred for fighting, but almost wholly for exhibition or practical purposes. Being a hardy race and having a good muscular development about the breast, they are used with good effect to cross on common stock, or on other pure-bred flocks. Game hens make the best of mothers, and are very courageous in defending their broods.

The Cornish Indian Games were some years ago introduced into this country from England, and while they at the time gave promise of becoming a popular market fowl, they, for some reason or other, are not much bred here at the present day. Their weight is: cock, nine pounds; hen, six and a half pounds. See Plate XII.

WHITE COCHIN BANTAMS.

The ninth class includes all Bantams other than Games. The breeds and varieties are numerous, but we illustrate only a few popular favorites in colored Plate XIII.

Bantams are bred mostly as pets for children, but are often profitably kept on city yards and village lots for their eggs and meat. For this service the Seabrights are an old and popular breed. For show purposes Bantams are bred down as small as possible, matured male specimens weighing only twenty-six to thirty ounces, and even less.

There are other breeds having decided merit not yet included in the list of fowls as given in the "Standard." Among these are the Sherwoods, said to have had their origin in Virginia from a White Game-Light-Brahma cross. They are a large, white, close-feathered fowl, combining the excellence of both parents.

SILVER-LACED WYANDOTTES.

The White Wonders are large, white, rose-combed fowls with yellow beak, shanks and skin, and have attained considerable popularity for their practical qualities. They have a reputation as winter layers.

The Argonauts are the result of an attempt by a noted New England breeder to produce a fowl having superior laying and table qualities, combined with good shape and pleasing color. The outcome is a chicken with pea-comb, buff color, and shape suggestive of Games, Plymouth Rocks and Dorkings.

There are also Rumpless fowls, having no tails to speak of; Frizzles, with feathers turned back towards the head; Silkies, with feathers so fine as to resemble animal fur; and Creepers, with legs so short that it is only by courtesy that they may be said to walk. These are kept mostly as curiosities.

The well-fed pullet is an early layer.

The swill barrel may become a chicken trap unless provided with a lid.

The wagon house makes a poor hennery. The cow shed and sheep pen are little better.

To break up a broody hen, shut her in the coop the first night you find her on the nest. The longer she sits the more "set" in her ways she becomes.

Chain the dog in the poultry yard at night. Prowlers will catch his scent and keep away.

Darkened nests will do much toward preventing the egg-eating habit. Use plenty of China nest-eggs. Let a few lie on the floor.

A good scarecrow, scarehawk and scarecat is a good gun in the hands of a good marksman.

Set the fodder cutter and crusher to cut fine, and run an armful of cornstalks through it. Scatter a bushel basketful every day on the floor of the poultry house.

GOOD MORNING!

PLATE XI.

SILVER GRAY DORKINGS.

Chapter XII.

TURKEYS AND GUINEA-FOWLS.

Plow up your dogs and plant turkeys.—Joaquin Miller.

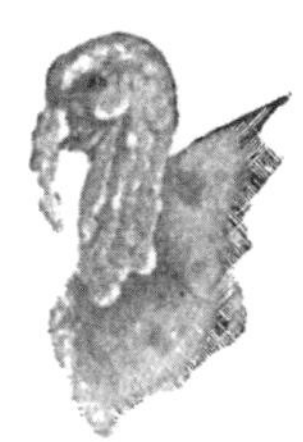

This noble bird, next to the chicken in importance among the denizens of the poultry yard, is a native of North America, and is found in a wild state from Mexico to Canada, east of the Rocky Mountains.

It is supposed that the wild turkey of Mexico is the parent stock from which our domesticated bird is derived.

Years ago the farm-yard flock was a somewhat variegated lot, but by skilful mating modern breeders have fixed certain characteristics of color and size so that we now have six quite distinct varieties, recognized and described in the "Standard of Perfection." The names of these, with the standard weight of adult birds, male and female, are the Bronze, thirty-five and twenty pounds; Narragansett, thirty-two and twenty-two pounds; Buff, twenty-seven and eighteen pounds; Slate, twenty-seven and eighteen pounds; White, twenty-six and sixteen pounds; Black, twenty-seven and eighteen pounds.

The weights above named are only reached, as a rule, by birds that are two years old or over. Sometimes they are exceeded even by younger specimens. In 1866, a Connecticut woman sent to President Johnson a gobbler, not quite two years old, that tipped the beam at forty-seven pounds.

Of the six varieties the Bronze, illustrated in Plate XIV, is the latest introduction. This originated by crossing the common with the Northern wild turkey. In plumage the Bronze resembles closely its wild parent, but the color is more brilliant. The lustre is like burnished gold in the sunlight, and it is almost an impossibility to properly reproduce it on paper or canvas.

WHITE HOLLAND TURKEYS.

The prevailing color of the Narragansett is a mixture of black and white, over which, in the sunlight, is seen a beautiful greenish-bronze lustre. The plumage of the Slate turkey is a grayish-blue. The White, or White Holland, is pure white, except the beard of the male, which is deep black. The red wattles, black tuft on the breast and the snow-white plumage of the rest of the bird make a striking contrast. A photograph from life of a pair of these birds is given on the opposite page.

The breeding of turkeys is more difficult than the breeding of chickens, because of the difference in the nature and habits of the birds. The turkey is not as thoroughly domesticated as the chicken, having been under the controlling influence of man but a comparatively short time and still retaining many of its wild traits. Their love of freedom, their roving habits and their shyness all indicate their recent introduction from the forest to the domestic state.

Young turkeys or poults, as they are called, are generally regarded as very tender until they reach the age of ten or twelve weeks. This is partly due to the unwise treatment of the breeding stock during the winter and early spring.

In the domestic state, turkeys pass the winter months in comparative inactivity. During this time they are fed principally on corn. When the breeding season arrives they are in prime condition for the table—fat and glossy, but are lacking in the vigor so essential for producing strong and healthy progeny. To this state of things may be attributed much of the weakness supposed to belong to them by nature.

As soon as the surplus stock has all been sent to market, the birds intended for breeding should be fed less corn and more muscle and bone-making food. One-third of their grain ration should consist of oats, and one-third of wheat, and the other third of corn, or corn and buckwheat. They are fond of cabbage, apples or any raw vegetables, and breeding stock should be well supplied with food of this kind. As the laying season approaches they should have nitrogenous food in the form of ground raw meat and bone or meat-meal, the former fed alone and the latter mixed in a mash of bran and corn meal.

When chickens and turkeys run together and are fed together the former will get at least two grains to the latter's one. For this reason, for fattening turkeys as well as for breeding stock, it is advisable to have troughs so made that the turkeys can feed at pleasure without interference from chickens. The illustration represents a cheap and handy feeder. It is made of six-inch fence slats nailed together for a trough and elevated to such a height that the other poultry cannot reach it. The end pieces and the lid are made of a foot-wide board, the lid being four or five inches above the trough. Slats at the bottom of end pieces give it stability.

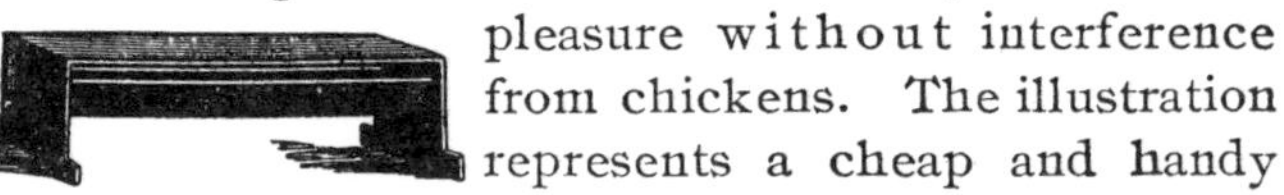

TURKEY TROUGH.

For breeders it is best to select hens two or three years old. If hens of the previous year are used they should be from the early broods. An early-hatched yearling male should be mated with old hens. When yearling hens are selected it is better to mate them with a two-year old gobbler. Young and undersized birds should in no case be used. Large, heavy toms should never be mated with small hens. One male is sufficient for five to ten females.

The turkey hen begins to lay in March or April, according to season and latitude. Her marked traits at this time are great shyness and secretiveness. She will seldom deposit her eggs in houses or nests where hens lay, but will choose rather a secluded fence corner or a bush, or bunch of weeds, or briars at some distance from the premises.

Before the laying season begins the poultry keeper should provide the hens with suitable nests not far from the buildings. This may be done by setting a few boards or an old door against a fence corner and throwing a bunch of hay under it, or by laying barrels or boxes on the ground in some secluded spot and putting a little hay in them. By a little strategy they may be induced to locate near by, and thus save the keeper much labor in looking after them and their broods.

As fast as the eggs are laid they should be removed from the nest, placed in a basket or box lined with woolen, and turned every two days. A nest-egg should always be left in the nest. By removing the eggs in this manner the hens will not become broody so soon and will lay a greater number. When the hens become broody, if there are more eggs than they can cover, set the rest at the same time under chicken hens, and when they hatch, which will be in thirty or thirty-one days, put all the poults with the turkey hens to brood and rear.

When the hens are tame and can be handled the young birds may be removed from the nest to the house, as they are hatched, until the whole brood is out, and then returned the night before the brood is put into the coop.

During the period of incubation the hen will

require nothing but corn and water and freedom from molestation. While the young are hatching feed and drink should be placed before her on the nest.

The poults require nothing to eat for twenty-four hours, and need not be fed until placed in the coop.

A familiar sight wherever turkeys are reared is the coop and yard made of foot-wide boards, here shown. For the first three days the mother should be kept in the coop, but after this may have her liberty. She will not go far away while her flock is confined. The pen should be located on well-drained ground, where there is short and tender grass. In the absence of grass in the runs finely chopped onions, lettuce or other vegetables should be supplied.

The diet of poults need not differ greatly from that of chicks. Hard-boiled eggs, so generally prescribed, may be safely left out of their bill of fare. Dry bread soaked in sweet milk is good for the first week. This may be given three times a day, and a little oat meal, finely cracked wheat or corn be kept where they can peck at it when so inclined. Ten young turkeys are killed by kindess in overfeeding for every one injured by starvation. It is not necessary to feed every two hours, as it is sometimes enjoined. It is more in accordance with nature to furnish them with food in such a manner that they cannot gorge themselves quickly, but will be compelled to peck a little at a time and often. Wet and sloppy food and fermented messes should be scrupulously avoided. Cottage cheese, made by scalding clabbered milk and pressing out the whey, makes a wholesome side-dish, and so does a custard of egg and milk mixed with bran and corn meal. Grit and water should be supplied from the beginning, as both are essential to health.

When the poults are able to hop out of their board pen they are strong enough to follow their mother. But as dampness is particularly injurious until they are ten or twelve weeks old, they should not be let out of the coop in the morning until the dew is off the grass, and it is always well to get them under shelter when a shower comes. Eternal vigilance is the price of sound and healthy turkeys at this early stage of their existence. If overtaken in a storm it is sometimes necessary to bring the little fellows in the house and dry them by the fire. As soon as they feather out and "shoot the red," as it is said when the red appears in their faces, they take on new vitality and can stand more hardships than chicks.

After this time they may be allowed to forage at pleasure. With a suitable range they will be able to gather in the fields and woods the greater part of their living. It is always prudent, however, to feed them twice a day, supplying them a light meal in the morning early and giving them all they will eat when they return at night. By taking care to feed them regularly in this manner they may be trained to come home every evening instead of perching on the fences out in the fields, or in the woods. But as "turkeys will be turkeys" now and then, and remain away from the premises, they should be hunted up the very first time their absence is noticed and driven home and fed.

If located near neighbors who also have flocks, the young poults should be marked with marking punches in the web of the foot. If the neighbors will agree to have different marks it will be an easy matter, if the birds get together, for each one to pick out his own.

In the fall when the harvest fields are gleaned, the grasshopper crop gathered in and insects become scarce, the birds are well-grown and lusty. The corn fields are now their favorite haunts and they are inclined to linger longer around the farm yard, and are eager for anything in the way of eatables their owner has to offer.

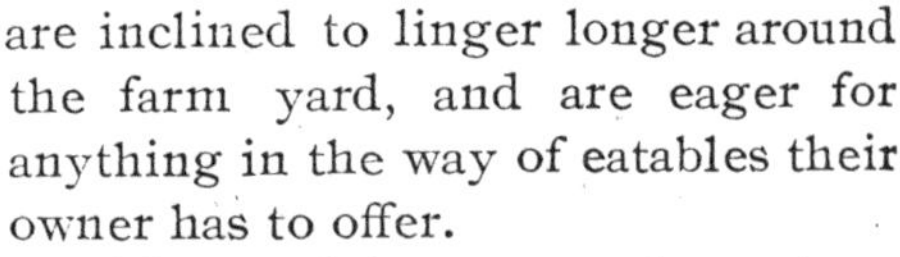

Thanksgiving comes along about this time and the first installment of the flock should be prepared for market and one of the best of the lot reserved for the farmer's own table. The illustration represents one of the flock the day after Thanksgiving. He is laughing all over his face now; perhaps Christmas day he will wear a different expression.

GUINEA-FOWLS.

The Guinea is closely related to the turkey and was originally brought from Guinea, on the West African coast, where it is still found in a wild state.

Their peculiar cry when alarmed will scare hawks and crows in the day-time. At night they are light sleepers and when aroused by thieves or other marauders their noise will arouse the neighborhood. They are great rovers and foragers, destroying many insects and weed seed, but doing little damage to crops. For making a gamey pot-pie no other domestic fowl equals the guinea. They lay many small but rich eggs and have a habit of secreting their nests in the fields and along fences, seldom ever laying near the farm buildings. In the hennery they are pugnacious and abusive toward other fowls, and their unceasing chatter is annoying to some people. Their good traits overbalance their bad ones and a few should be in every farm-yard.

One male is sufficient for a flock of six to ten females. It is well to set the eggs under a chicken hen. Reared in this way they are more domestic. They will follow the mother-hen, to her great annoy-

A FLOCK OF PEARL GUINEAS.

ance, until they are full-grown. The young are quite hardy and require no special treatment or care different from chickens or turkeys. The plumage of the Pearl Guinea, the most common variety, is a groundwork of blue sprinkled with pearl dots of white. The males usually have some white on their breasts, have larger wattles and larger bodies than the females. The Whites differ only in color, and are probably a sport of the Pearl.

THE PEA-FOWL.

The most gorgeous in plumage of all our domestic birds is a native of Southern Asia and the Malay Archipelago. They are kept for ornamental purposes only, being of no practical value. One pair is enough for a whole neighborhood, as by their shrill cry at night they can awaken everybody within a radius of half a mile. The mother-hen usually steals her nest and brings up her brood without any assistance.

PLATE XII.

INDIAN GAMES.

CHAPTER XIII.

DUCKS.

Ducks and rats do not thrive in the same house.—Tim's Wife.
A duck's appetite is as big as the feed bin.—Tim.

The domestic duck is believed to be a descendant of the Wild Maliard, the most common and numerous of the wild species. Four varieties are recognized in the "Standard of Perfection"—the Rouen, Cayuga, Aylesbury and Pekin.

Rouens are regarded as a French breed and appear to be the Mallard domesticated and enlarged by selection and breeding. The pair seen in the foreground in colored Plate XV, fairly represent the shape and beautiful plumage in which this variety is clothed. The standard weights of adult birds, male and female, are nine and eight pounds respectively. They are hardy and are prolific layers of large greenish eggs.

The Cayuga is an American variety, jet-black in plumage, supposed to have originated near Lake Cayuga, New York, from a cross of Mallard and the Wild Black, or Buenos Ayres duck. The standard weights for these are eight and seven pounds respectively.

The Aylesbury is the favorite English variety. The plumage is a pure "dead-white" throughout, the beak a pale flesh-color, and the shanks a light orange. The standard weight is the same as for Rouens.

Another duck sometimes seen in the farm-yard is the Muscovy. This belongs to a different genus from the varieties already described, and is a descendant of

A LONG ISLAND DUCK FARM.

the wild Musk Duck of South America. There are two varieties, the colored and the white. The latter is shown in the background in colored Plate XV, which well illustrates the peculiar shape and appearance of this duck, which differ decidedly from that of the common varieties. They will breed with other ducks, but the hybrids are mules, or sterile. While kept mostly as curiosities, or for ornamental purposes, the crosses are said to make excellent market poultry.

The Pekin is an Asiatic variety having been first imported from Pekin, China, in 1873. The plumage is white with a creamy-yellow shading, the feathers being downy and fluffy like Asiatic chickens. While the "Standard" gives their weights as a pound lighter than Rouens or Aylesburys, they are commonly regarded as a larger duck than either.

The introduction of the Pekins to this country gave a new impetus to duck breeding, and many persons have entered into it on an extensive scale. While they are prolific layers of large eggs, mostly white-shelled, they are also the great market duck. Their bills and shanks are a deep orange-yellow and their skin also is yellow. As the plumage is white and the pinfeathers leave no stain on the flesh, they make the finest dressed carcass of any variety.

The keeping of ducks for eggs is a profitable part of the duck business, when rightly conducted, and the keeper is within easy access to a city market. During the early spring months duck eggs bring higher prices than hen eggs, and it is at this season that ducks are most prolific. To obtain the best results from eggs the laying ducks should be hatched the latter part of the breeding season, in June and July. The spring-hatched will grow larger and will make better breeding stock, but with proper care these late broods will lay as soon as the severity of winter is over, as soon, in fact, as the early-hatched.

and will not require to be fed during March, April and May. The proper feed for such ducks, to induce early and prolific laying, is well illustrated by the practice of a successful breeder who commonly winters about five hundred. He feeds them on equal parts of boiled turnips, wheat, bran, and corn meal, with a little—say ten per cent.—of ground beef scraps thrown in. This is mixed thoroughly together while the turnips are hot, and constitute the entire feed during the winter and spring. About the first of January or a little later, when they begin to lay, the proportion of bran and meat scraps should be increased.

This mess is fed morning and evening, and at noon they have a light meal of dry food composed of equal parts of cracked corn, oats and wheat.

Ducks kept expressly for market eggs require no drakes with them, which is one of the points in favor of this part of the business. As soon as the price drops and the egg supply begins to run low the layers should be sent to market.

When large numbers are kept, either for laying or breeding, large houses properly constructed are required. The character of these houses will be determined by the climate and other circumstances. Where the winters are mild and snow seldom tarries long on the ground long open sheds will suffice; but where the winters are long and severe and snow lingers, large, storm-proof houses are needed. While ducks are hardy and can endure more cold and wet weather than chickens, when early laying is the object sought the layers must be shielded from the severity of the weather. James Rankin, in his excellent treatise on

Duck Culture, described the house in which he keeps his breeding ducks through the winter as covering fifteen by two hundred feet floor space, having five-foot posts in the rear and four-foot posts in front, and an uneven double roof, the short slant being in the rear. There is a walk through the rear, three and one-half feet wide. The building is divided every twenty-four feet into pens, in each of which forty ducks are wintered. The partitions are but two feet high. The walk is separated from the pens by lath three inches apart, to allow the birds to feed and drink from troughs placed in the walk. This arrangement enables

ONE OF JAMES RANKIN'S DUCK HOUSES.

an attendant to feed and water the whole houseful in a few minutes, a wheelbarrow or truck being used for carrying supplies; it also prevents waste of feed or fouling of the feed or water. Only ten feet of this slat partition along the walk in each pen is used for feed, and four feet is made movable so that the attendant can enter with barrow to clean out the pens. The other ten feet along the walk is lined with the nests, which are fifteen inches square, the back and division boards being a foot high and the board next to the pen but four inches, or just high enough to keep the nest material in. This latter consists of cut straw or hay, which is kept dry and clean, thus preventing the eggs from becoming soiled and stained. With such a house there should be either joined or situated near-by a feed and cook-room containing bins, a root-cutter and a capacious boiler. The front of the building is one-third glass. From the front the yards extend one hundred feet, making each one twenty-four by one hundred feet. Experience has proved that free range and water are not essential to success in keeping ducks, especially Pekin ducks, for laying or breeding.

Ducks as a part of the farm poultry should be kept apart, as much as possible, from the chickens, and away from the barnyard and farm-yard and out of sight of the dooryard. With the chickens they foul the drinking water and the food and their feathers become soiled in the hen-house. In the barnyard they are liable to be trampled by the stock, and they are too filthy to be tolerated in the farm sheds, or on the grass of the lawn. They should have houses, shelters and yards of their own in all cases. These need not be expensive. The houses may be low, and no fence for Pekin ducks need be over two feet high.

An excellent shelter for a farm flock is a shed, one-half of which is open and the other half closed. The open half should have a movable slat fence or gate for use when it is desirable to confine the flock. If they have free range it is necessary to confine them to a house or yard for two or three hours after daylight during the laying season, otherwise they will drop their eggs in the fields and meadows, or along the streams, and many will be lost.

A convenient form of duck-house is here shown. As ducks are humble-minded creatures they do not require a lofty building, and therefore one for their accommodation may consist principally of roof. It is a movable house six by ten feet, set on plank runners fifteen inches wide. This structure, set on a well-drained site, bedded with short hay or straw and moved occasionally, will serve as headquarters for a flock of ten to twenty-five.

Breeding ducks should be carefully selected for their size and typical shape, and only mature birds should be used. An active yearling drake may be allowed for each five or six ducks. As the drakes are not so pugnacious as cocks, flocks may contain several of them without danger of their injuring one another.

As a general thing it is better to hatch duck eggs under hens than under ducks. The period of incubation for duck eggs is twenty-eight days, and the temperature required is the same as for hen eggs. They have strong vitality and are easy to hatch either in the natural way or artificially.

Ducklings when hatched are animated balls of down, seldom quiet and never so happy as when eating or dabbling in water. They do not require so much warmth from the mother and do not need to be hovered so much as chicks. Hence, it is safe to put thirty to forty with a single hen. More also can be put in a single flock in a brooder than of chicks.

While ducklings will take to the water as soon as hatched, they do better if not allowed to swim until they are four weeks old, and should not be allowed to enter ponds or streams until they have their first feathers. Thousands of ducklings die yearly from cramps and convulsions, because they are allowed to enter the water too young or too early in the season while the water is cold. Cold spring water even in summer is fatal to them.

For the first ten days ducklings, with hens, do best in small yards, like those described for confining young turkeys. The coop should have a board bottom, to prevent the hen mother from scratching earth over her downy brood. All the water they need is enough to drink and to dip their heads into, to wash out their nostrils and eyes. It is difficult for a duck to eat without the frequent use of water. A duckling will drink about one hundred times, more or less, while eating a single meal. The water vessels, therefore, should be close to the feeding trough, but so arranged that they cannot get in them with their feet or dip their heads in deep enough to throw water over their backs.

AN UNNATURAL FAMILY.

Healthy ducklings have a voracious appetite and will eat whatever is set before them. Dry bread soaked in milk is excellent food for the first two days. In passing it may be said that it is not advisable to give ducklings milk to drink; it should always be used for mixing their feed. They will get it on their down and in their eyes, and thus not only spoil their good looks but injure their health. After the first few meals of bread and milk, equal parts of corn meal and wheat bran, wet with milk or water, may be fed. A little fine-ground meat scraps, or meat-meal, should be added. After ten days every other meal may consist of cracked corn and wheat. Care should be taken to have all their food crumbly rather than doughy or sticky. At first they should be fed every two hours, but at the end of a week they can get along with four meals a day. Like all other birds they need grit as soon as hatched. To supply this at first it is a good plan to sprinkle a little coarse sand on the feeding board or in their feeding trough. When a little older put the grit in the bottom of the drinking vessel.

The yard in which the ducklings are placed should contain short grass, but if it does not, green food in some form must be supplied regularly and bountifully. Lettuce, beet tops, cabbage, green clover, or green corn cut fine, will be greedily devoured.

While they are hearty eaters they are, for this reason, rapid growers and will increase in weight about twice as fast as chickens. They are usually slaughtered when from seven to ten weeks old.

In warm weather it is important to have some shelters for ducks and ducklings confined in yards. If the latter contain no trees, vines or bushes, temporary shelter of boards, brush or canvas must be provided.

Temporary yards may be made for ducklings by the use of wire netting two feet wide, stapled loosely to stakes driven into the ground. Such a fence is easily moved by pulling up the stakes with the wire on them and rolling all up together.

The Swan (Cygnus), first cousin of the duck and the goose, is frequently referred to as the type of graceful beauty in outline and motion. There are numerous varieties, nearly all of them found in a wild state. Formerly the bird was served at feasts on special occasions, but it is now kept in private and public parks solely for ornamental purposes.

DUCK NOTES.

Quack! Quack!! Quack!!!

Harvest-hatched ducks make good spring layers.

Ducklings will kill rose-bugs, and rose-bugs in large doses will kill ducklings.

Ducks being water-fowl are warm-blooded and like water, but appreciate a dry floor to roost on. Having a water-tight roof the floor can be kept in proper order with cut straw or leaves and dry earth. The litter should be *short*.

The sex of ducks can easily be distinguished by the quack. The voice of the male is pitched in a high key and that of the female in a low key; the male has a larger head and thicker neck and when in full feather one of the tail feathers is curled backward.

White clover sod does not make a good pasture for ducklings. Bees like white clover as well as ducklings, and consequently the three get badly mixed up. The bee stings as he goes down the duckling's throat on a clover head, and the career of the bee and duckling both come to a sudden termination.

SINGLE FILE.

PLATE XIII.

GOLDEN SEABRIGHT. BLACK-TAILED JAPANESE. ROSE-COMBED BLACK. BLACK-BREASTED RED GAME. BUFF COCHIN.

BANTAMS

CHAPTER XIV.

GEESE.

It is a silly goose that comes to a fox's sermon.
The goose that has a good gander cackles loudly.
—Danish Proverb.

Farmers who have rough, marshy land, may with little extra expense and labor add to their incomes by stocking it with geese.

Our domestic goose has descended, it is said, from the wild greylag goose of Northern Europe. The common gray and white geese of the American farm-yard need no description, since they are well known everywhere. The Toulouse, a large, gray variety, has come to us by way of England. Their shape and color are seen in the foreground of colored Plate XVI. The difference of the sexes may be plainly seen by observing the head and neck of each bird. The gander has a larger head and thicker neck than the goose. But it will be noted that the abdomen of the latter is heavier and closer to the ground. The standard weight for adult Toulouse is forty pounds per pair. They sometimes attain greater weights than this, but not until three or more years of age.

There is a large, white, pure-bred variety called the Embden or Bremen, so named from two towns in Hanover, in northeastern Germany, where they are supposed to have originated. The Embden has pure white plumage, prominent blue eyes, a flesh-colored bill and bright orange legs. The weight is about the same as that of the Toulouse.

Chinese geese, or swan geese, belong to another species, and are at once recognized by a peculiar knob or protuberance at the base of their bills and by their long, swan-like necks. There are two varieties, the White and the Brown. The latter is shown in the background of colored Plate XVI. The standard weight of these is twenty-eight pounds per pair. African Geese, recognized in the "Standard," belong to the same species and are similar to the Brown China, but heavier.

The American wild, or Canada goose, belongs to a different species from either of the above, and will not produce a fertile cross. It has never become thoroughly domesticated and does not breed readily in confinement.

Geese are long-lived, and the females may be kept for eight or ten years, but the ganders become pugnacious and less virile after they are three years old. It is best, therefore, to mate old geese with young ganders, allowing one male to two or three females. The geese agree better if selected from the same flock. To avoid in-breeding, select the male from a different flock. Geese incline to go in families and are very jealous of their mates. For this reason, when there is more than one flock or family, it is prudent to have separate sheds for each one, and if possible, separate runs.

In northern latitudes it is not well to feed breeding geese too generously in the winter and start them to laying early. Goslings and green grass should appear about the same time. But conditions being right, the earlier goslings hatch the better. About the first of February in the Middle States the forcing may begin, the breeders being fed in a manner similar to that recommended for breeding ducks. During the winter cut hay, ensilage and a little corn with refuse vegetables will sustain them, but now they should have nitrogenous food, like bran, shorts and meat scraps fed with cooked vegetables.

The goose will lay two litters of twelve to fifteen eggs each. If well fed this number may be increased. The China goose is said to sometimes lay from fifty to sixty in a season. To get the eggs all hatched as soon as possible the first laid may be hatched under hens, allowing each hen to incubate from five to seven eggs. When the goose has finished her first laying and becomes broody she may be confined for a few days and be well fed. When her brooding fever is over she will lay again and may be permitted to hatch and take care of the second litter.

The period of incubation is the same as that of ducks, twenty-eight to thirty days.

Goslings are hardy, but should, like ducklings, be kept in a pen for two or three weeks and allowed only water enough to drink. Since goslings are regarded as a great delicacy by snapping turtles, minks and other varmints, it is well to keep them from infested ponds and guarded at night in sheds enclosed with netting. The later hatches, left to run with the mother-goose, will require less attention and care, but yet it is advisable to confine the flock in a yard for a week or ten days.

When the goslings are to be sold in the Christmas markets, or late in the year for breeders, they will not need to be supplied with food if they have suitable pasture grounds, except a light meal of grain morning and night. It is best to feed them in this manner to induce them to return home every night.

There is a demand for "green goose" in midsummer and many prepare their early goslings for this market. With this end in view they are fed all

EMBDEN GEESE.

they will eat until the flight-feathers grow out as far as the root of the tail, then they are enclosed in a pen. This must be in a dry situation where there is no water or mud. A yard fifty feet square with shade in it will hold seventy-five goslings. Treat them gently, since they are timid creatures and will not fatten if roughly handled or frightened. Have a large boiler holding a barrel or more, fill with water and stir in the boiling water, meal and twenty-five pounds of meat scraps to the barrel. Mix till as thick as can be stirred. Season with a little salt. Feed all they will eat of this and give only enough water to drink. Furnish gravel and put in the enclosure some rotten wood. In seventeen to twenty days they will be ready to slaughter. They should be in market before the fourth of July.

One source of profit from geese is the feathers, which are always in demand at good prices. These are obtained not only from the slaughtered birds but also from the live ones. When done with discretion the practice of plucking is not so cruel as it might at first sight appear. Four times a year is often enough to perform this operation. Never pick when laying, nor in cold weather, and pick only when the feathers are "ripe." This ripeness is detected by the experienced eye by the dull, dead color of the plumage, and in Pekin ducks by the absence of the yellowish tinge. To test them pluck a few from the breast. If they come easy and are dry at the quill end they are "ripe," if the least bit moist or bloody do not pick any more. In picking, take only a small pinch of feathers in the fingers at a time, and make a quick downward jerk from tail to neck. Remove only a little of the down. Never remove from a live bird the cushion or bolster of coarse feathers along the side, that supports the wing.

The goslings may be picked as soon as they are full feathered. An experienced geese breeder thus describes his plan of making the most out of the feather crop: I like my geese to hatch out about the last of April. At that time I pick the ganders of the flock, the geese having lined their nest with feathers they are not in condition to be plucked. About the first of June the ganders are full feathered again and the geese are ready too, as you will begin to find loose feathers where they stay over night. Then in about seven weeks the goslings are ready to be plucked with the old ones. Don't take the feathers off too bare, as the sun is hot at this season. By the last of September you will get a fine lot of good feathers again. If you keep the geese for the holiday market they are again ready in early November, but if the nights are cold drive them up and give shelter. They will soon feather at this time of year, and at killing time you will get the finest crop of the year.

Fasten them up in a stable having plenty of clean straw under them for half a day before you begin to pluck the feathers, then they will be dry and clean. Take a narrow strip of muslin, tie their feet together, lay them on their backs, tuck their wings under them, let an assistant take hold of the head, and as soon as they are done struggling begin to pluck.

There are no disease germs in fresh eggs.

Poultry products sell for cash, and can be sold at any time. Two important points in favor of the hen business.

In long houses, instead of an entry and tramway for carrying feed and water have an overhead track and suspend a platform car on which to carry buckets and boxes. Will be useful, also, in cleaning the house, carrying manure out and fresh gravel in.—Tim.

The crops of fowls should be empty when sent to market. The best way to secure this condition is not to feed for at least twelve hours before killing. If for any reason the crop be full after killing, make a cut two inches long through the thick skin on back of the neck, insert the finger in the incision, draw out the crop and cut it off. The mutilation will not be apparent.

PLATE XIV.

BRONZE TURKEYS.

Chapter XV.

PIGEONS FOR MARKET.

A bird in the loft is worth two in the pot-hunter's bag.

In a neighborhood where pigeons fly both peas and peace take wing.—Tim.

The old practice of fastening nest-boxes on the outside of building and allowing the occupants to range at will is not to be commended However made they present an unsightly appearance, and pigeons at liberty in a community are an intolerable nuisance.

It is better in every way to have a separate building for pigeons, and to have an outside fly of wire netting connected with it and thus to keep the birds confined at all seasons. This plan is especially recommended when any considerable number is kept.

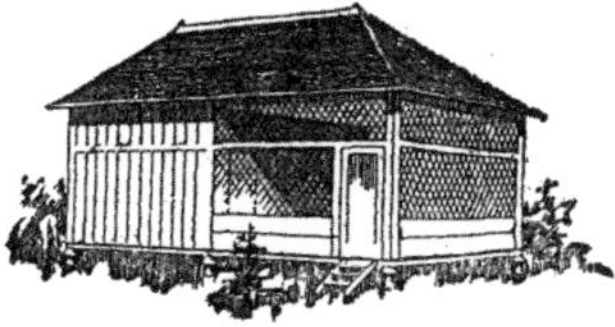

The accompanying illustration shows a loft with the breeding-room eight by sixteen feet and a cage or fly sixteen by sixteen feet that will accommodate twenty to forty pairs. In building it posts are set firmly in the ground, pans inverted over them to keep out rats and mice and the sills nailed to the posts.

For larger numbers the house shown on the next page illustrates a cheap and practical building. It is eight by thirty-two feet, but may be made any length desired. The front is ten feet high and the rear six feet. The roof, rear and end wall should be wind and rain-proof, but it is well to have a considerable portion of the front open, especially in summer. Netting with two-inch mesh will confine pigeons, but where the English sparrow abounds one-inch mesh is preferable The floor cf the loft may well be of earth, but should be dry.

The nest boxes in a loft should on no account be made in rows on permanent shelves and of a uniform appearance. Instead of regular rows of nests of one pattern use large soap boxes, starch boxes, irregular boxes, nail kegs or anything that will give individuality to the home of each pair. Do not nail these fast to the walls or beams, or set them on shelves in regular order, but hang them on hooks or screw-eyes so they can be easily taken down.

Figure 1 illustrates how a soap box may be transformed into a first-class home for a pair of breeders. A division board is placed in the middle and alighting boards at either end. Figure 2 shows a smaller box containing but a single nest, so made that no alighting board is needed and the roof sloped to prevent perching upon it. Two of these will be needed for each pair and should be placed adjacent. Nail kegs may be suspended by wire to beams or rafters and have the open end a little higher than the other, or a piece

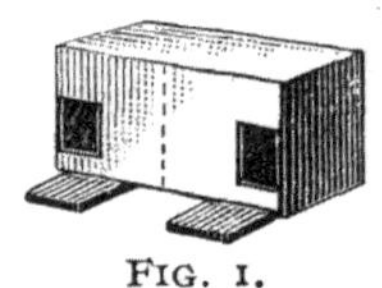

FIG. 1.

FIG. 2.

of the head of the open end left in, to keep the eggs and squabs from tumbling out.

The irregularity in shape and arrangement of nests may shock the fastidious, but will avoid contention and confusion among the birds, which frequently results in the loss of eggs and squabs.

POUTER.

For raising squabs for market it has been proved in late years that the common pigeon does not give the uniform, plump, attractive carcass that the market demands, and which is credited to the Homer variety. Some advise crosses with Runt and Dragoon, but it is generally conceded among squab growers that the Homer in its purity gives all the requirements of a squab to meet the demands of the most fastidious.

It is desirable to have breeders that raise squabs with light skin for they always bring the top price. The color of the skin is not controlled, as is popularly supposed, by the color of the feathers. Parents with white plumage may have dark squabs, and those as black as crows may produce squabs with fair skin.

A good plan to stock a loft is to buy enough mated birds to fill it one-fourth full, and raise enough from these to make up the complement, selecting the young from the parents that prove to be prolific, and raise the largest and whitest squabs. As mated birds are not always obtainable the next best plan is to buy squabs just able to fly. A good time to buy is in June, July and August, when squabs are low in price. These birds will pass their moult

JACOBIN.

and begin to breed in the following winter and spring.

Pigeons breed in pairs, and when once mated remain faithful to each other unless the union is broken by death or by the coquetry and intrigue of unmated birds. The latter are sure to make mischief and care should be taken to exclude them, or to remove them from the loft when discovered. It is always best to mate pigeons, that are not known to be already mated, pair by pair, before turning them into the loft. This may be done by placing the couple in a coop or cage alone for two or three days. The novice may attempt to mate two of the same sex. If both be males, the cooing and strutting and fighting will make the mistake evident. If both be females, there will be no love-making, but may be some quarreling. How to distinguish the sexes frequently puzzles experts. The experienced eye can generally detect the masculine or feminine features of a bird, and will name the sex nine times out of ten. There is no way to get this experience except by long and careful observation. The female is smaller, as a rule, than the male, and has a feminine look about the head and neck, the eyes being milder, the head narrower and the neck more slender than the corresponding parts of the cock.

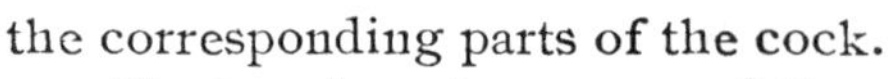

The hen lays two eggs and then both birds assist in hatching them. The hen sits all night and a part of the day: the cock sits the balance of the time. Both assist in feeding the squabs. If the hen lays again before the first brood are out of the nest the cock will

FANTAIL.

usually take entire charge of the young besides doing his share of incubations. The two eggs will usually hatch one male and one female.

The natural food of pigeons is grain and the seeds of grasses. They are fond of millet, clover seed and peas, and if allowed to fly when these crops are sown will prove very destructive. Hemp seed is to pigeons what candy is to children. A little may be given them on entering the loft to tame them.

For a steady diet the following is commended: two parts whole corn, two parts wheat and one part buckwheat, all to be old, sound grain. Screenings to be economical should be purchased for one-fifth the price of good wheat. New grain is not good for the squabs. The corn should be a variety having small grains and should in no case be cracked.

In order to supply feed for the very young squabs it is well to keep equal parts of bran and corn meal in self-feeding hoppers always before the breeders. Experience has proved that the old ones feed with greater regularity and fatten their young better when the whole grain is supplied at regular hours, three times a day, all they will eat up clean. They will not eat grain that is fouled, if they can avoid it, and should not be compelled to do so.

TUMBLER.

For side dishes they should have ground oyster shell in a box or barrel lid where they can help themselves, a lump of rock salt and a bit of salt codfish tacked to the side of the loft by several nails, so they can peck at it, but not tear it down.

The floor of the loft should be kept reasonably clean and be strewn occasionally with fresh sand and gravel. Red gravel is the best, as it contains iron, the oxide of iron giving it its peculiar color. Pigeons will peck at clay and coal ashes, and also at weeds and grasses. They use these substances, probably, for medicinal purposes, as dogs eat grass and cats eat catnip.

Pigeons drink a great deal of water, and it is important that it should be kept clean. Open vessels should never be used in a loft, unless a stream of pure water can be kept running through them. A wire cage like the cut, open at the bottom and closed on top, set over a basin, makes a handy arrangement. Stone or earthen self-feeding fountains, such as are used for fowls, are good.

A daily bath in summer, and twice a week in winter, is essential to the comfort and health of the flock. Wide, shallow milk pans answer very well for bath tubs. These may be set out in the fly filled with water, and allowed to remain an hour or two and then emptied.

An open feed-trough is quite as objectionable as open water vessels. The feed in them becomes foul and much of it is wasted. The self-feeding hopper shown in the accompanying illustration is one of the best that can be found. These hoppers can be made of starch or soap boxes, by any one handy with tools. The lid should be broad enough to cover completely the feed trough at the sides.

and these troughs should be just broad enough to allow the birds to feed without permitting them to get in with their feet.

Pigeon eggs hatch in sixteen or eighteen days. After the first few days the young ones grow with wonderful rapidity, if the parents are supplied with proper food and do their duty. In from four to six weeks the squabs are old enough to kill. Some develop so much more rapidly than others that no fixed date can be given at which it may be said they are of the right age to be in the best condition to sell. When this period is reached the neck feathers have passed the pin-feather stage, and the tail is usually about three inches long, but the bird is still unable to fly. When they begin to fly they are too "hard." as dealers say, and when the skin of the crop and of the abdominal pouch is thin and transparent and these parts are full and the breast undeveloped, the dealers complain that they are too "soft." It often happens that one of a pair—it is usually the male—is ready for market a week before its mate. By marketing the larger and leaving the smaller one to be nursed by the parents, it will be ready to go with the next lot.

Squabs are killed and dressed just like chickens, by bleeding in the mouth and picking dry. They are in the best condition for killing in the morning before the old ones give them their breakfast.

After killing and dressing they may be tied in pairs, or in half dozens, and put into cold water, or packed on ice until sent to market.

Where breeders are a long distance from market

it is better to send squabs in crates alive. In this case they must be old enough to fly, or, at least, old enough to feed themselves.

There should be a weekly slaughter on a fixed date in the week. On these occasions every nest should be examined so that no bird that is old enough may be overlooked or get away.

A well-managed flock will raise, on an average, five pairs of squabs annually for every pair of birds it contains. It is not safe to base calculations for profit on a greater increase than this, although it is quite possible.

Prices vary with the season, rising in the winter and spring and falling in summer. Near the large eastern markets it is safe to reckon on an average of forty cents a pair. This will make the returns from one pair of breeders $2.40 a year. During this time the parents and their progeny will consume food worth at least $1.50. This will leave a balance to their credit of ninety cents. The droppings of a pair of pigeons in confinement are worth ten cents a year, which will make the profit, not counting labor, an even dollar. It is possible to do better than this and possible also to do worse.

PLATE XV.

ROUEN AND MUSCOVY DUCKS.

CHAPTER XVI.

FATTENING AND MARKETING CHICKENS.

Well-fattened and cleanly dressed poultry is half sold.
The market is never overstocked with strictly fresh eggs.
—Tim.

It is a waste of time and food to sell any but well-fed, well-conditioned and well-dressed poultry. Sound yellow corn is the best grain for fattening purposes. The more of it fowls can be induced to eat and digest, the quicker they will fatten. Whatever else is furnished should be given as a condiment to aid in the assimilation of the corn. Two of the three meals of fattening fowls should consist of corn meal mixed with milk and seasoned with salt. For the noon meal whole corn and wheat with a little vegetable food of some kind and a little meat may be given for a change. Clean water, plenty of sharp, gritty gravel and a box of granulated charcoal should be kept before them at all times. Food should not be permitted to lie before them but they should have at each meal all they will eat up clean, and every bird should have a chance and time to get his portion. Fowls will continue to improve just as long as they continue to eat with a relish. How long this will be depends much upon the skill of the feeder. From ten to fourteen days is the time usually allowed for fattening chickens. It is difficult to carry on the process longer in coops, but in small yards and under skilful hands it may be prolonged for a month. As a rule the operation can be most quickly and economically done in a properly made coop. Figure 1 illustrates one that is admirably adapted to the purpose. A portion of the front wall is cut away to show its interior. It is eight feet long, three feet wide and four feet high in front, two-and-a-half feet high in the rear, and set two feet from the ground.

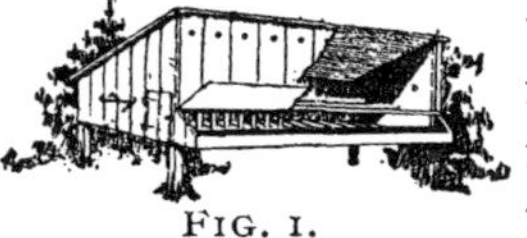

FIG. 1.

A pole is attached to a movable partition, which slides on slats. When it is desired to catch the fowls, by laying hold of the pole where it protrudes through the end the fowls are all drawn up close to the door. The bottom is made of slats. The feeding trough is six inches wide and four inches deep and has a lid.

When large numbers are to be prepared at one time a fattening coop is not available. But wherever it may be done the birds should be kept in a quiet and restful state. This will preclude the putting together those of different flocks and ages that are likely to fight and keep up a turmoil in the pen.

Ducks and ducklings do best in small pens or yards. The same may be said of goslings. Old geese will fatten while running at large. Water fowl need more vegetable food while fattening than do chickens. No poultry, however, should be fed green vegetables or grass for two days before being killed. Onions, turnips, cabbage, fish or other food having a pungent odor should not be fed during the fattening period.

Turkeys do not thrive well in confinement and can best be fitted for market while on the range, but special care should be taken for a month or six weeks to let them have all the fattening food they can be tempted to eat.

The caponizing or emasculation of male chickens may be mentioned here, as it pertains to their better preparation for market. The manner of performing the operation can best be learned under a skilled operator, but those who sell the necessary instruments send with them instructions from which, with practice, any one may become proficient in the art. The effect on chickens is the same as on animals, it makes the subject quieter in disposition and greatly improves the quality of the flesh. Capons, therefore, are easier to manage, easier to fatten and bring a better price than any other poultry except early broilers.

It is generally the later hatched cockerels that are caponized. The earlier ones pay best to sell as broilers or roasters. All hatched before the 1st of April can be marketed before the July drop in price occurs, or kept over for the early fall trade. The cockerels of the April, May and June broods are ready for the operation in from three to four months from hatching and will have ten months in which to grow for the capon market, which includes the period between the middle of January to the middle of April.

The breeds best adapted for capons are the medium and large ones or their crosses.

In preparing and dressing poultry for market the intelligent poultry keeper will seek to learn what the general market requires and what special requirement is made by the market to which he is about to ship. Dry-picked poultry sells best in all markets. By this manner of dressing the skin retains its color and the flesh its natural firmness. When scalded the skin turns blue, tears easily and peels off, giving the carcass an aged and uninviting appearance. It pays to dry-pick and when the art is learned it is a speedier method than scalding.

To dry-pick with ease and dispatch the bird should be hung up by the legs at a convenient height, and bled by making a cut across the back of the mouth, finishing by a deft thrust of the point of the knife into the spinal cord at the base of the brain. This paralyzes the bird, relaxes the muscles and loosens the feathers. This last thrust is acquired by practice and makes dry-picking easy and rapid.

Poultry for the New York and Philadelphia markets should be plucked clean. Capons should have the feathers of the head and neck, tips of wings and the tail left on. The first joint of the wings of ducks and turkeys is usually removed along with the feathers and retained by the farmer's wife or sold for dusters.

CAPONS FOR PHILADA. MARKET.

Boston must have its poultry "drawn"; that is, the entrails removed. Broilers need not be drawn. Ducks should have the tips of wings left on and the wings tied to the body, to retain the shape of the carcass.

Baltimore also requires poultry to be drawn.

Chicago wants its poultry dry-picked, with heads off and the skin drawn over the neck and tied, and the entrails removed.

In all cases when dressed poultry is sent to market undrawn, the crop should be entirely empty. This condition may be secured by not feeding them for twelve hours before killing.

Some markets demand yellow-flesh fowls, others prefer white, but all want plump, nicely fattened stock.

In packing poultry dry for shipment to market use clean barrels or boxes holding about two hundred pounds. Line the case or barrel with clean manilla paper, but use no packing. Place the poultry in breasts down and legs out straight, crowding them together closely so as to fill the entire space. Put paper over the top layer and fasten a cover of burlap over the barrel and slats over the case. Poultry can be shipped in this manner in cool weather. It must be thoroughly cooled before packing and all blood and stains wiped off.

For warm-weather shipments poultry must be packed in ice. For this purpose sugar barrels are commonly used. Holes are first bored in the bottom for drainage and a layer of broken ice put in the bottom. A layer of poultry is put on this ice, breast down, heads out and feet towards center. The layer of poultry complete, a layer of ice is put on and then a layer of poultry until the top is reached, when one or several large lumps are piled on top and a burlap cover over all.

The address of the consignee and the weight of the poultry should be placed conspicuously on the cover, along with the address of the consignor. When mixed lots are sent, if large enough, it is best to pack separately, or, if packed in one barrel, they should be grouped together and the weight of each noted on the cover.

All shipments should be made so as to be sure to reach the market before the close of the week and at least three days before a holiday.

For long-distance shipments poultry is usually cooped alive in crates or hampers made for the purpose of slats or of wire and splints. Different kinds of poultry and birds of different ages and sizes should not be crowded into one hamper or the smaller and weaker may be trampled to death by their stronger companions. To be sure of rapid transit it is safest to ship poultry by express, but as to this every one must be guided by circumstances.

Eggs are now nearly all shipped in crates having what are called pasteboard "fillers." The standard crate holds thirty dozen and the gift form of it that is sold with the eggs is popular with dealers. The producer of eggs who can ship once or twice every week to a dealer or grocer having good customers, and who will send only clean and strictly fresh eggs, can usually get a few cents above the market price. The vicious system of collecting and marketing eggs in vogue in this country is responsible, to some extent, for the low prices that prevail at certain seasons. The eggs are left in the nests a few days, then kept in the house for a week, then traded for groceries at the village store. In a week or two they are sent by the groceryman to the city and through a dealer are distributed to city grocers, finally reaching the consumer as "fresh country eggs."

A successful egg farmer who made money at the business always shipped his eggs in sealed crates to a dealer who had a gilt-edged trade and guaranteed every egg to be fresh and sound. The dealer sold them under this guarantee to customers who were willing to pay an advanced price for such stock. The result was satisfactory to all parties concerned.

A POOR FATTENING PROCESS.

If you have bought tarred felt to line the poultry house with to keep the flock warm, don't do it. Put it on the outside. Brighten up the inside with lime wash.

Keep wood ashes out of the hen house. A small portion may be mixed with the loam in the dust-box for medicinal purposes. Wood ashes bleach the shanks of fowls and when mixed with the droppings cause the ammonia to escape.

Notice with what pleasure a hen scratches among the forest leaves in summer. This is a hint to save the leaves to scatter on the floor of the poultry house in winter.

Corn meal fresh from the mill will quickly heat and spoil in warm weather if not looked after. Mix with bran and stir it up occasionally. If it becomes mouldy and caked throw it on the manure pile; do not feed to fowls.

To preserve eggs for family use, pack strictly fresh ones in fine salt, small end down, so they do not touch each other. When the box is full screw lid on and turn twice a week.

A person who formerly kept a large flock of laying hens and had an old-fashioned stationary boiler, put in whatever vegetables and meat he had to cook about the middle of the afternoon, filled the boiler nearly full of water and started the fire. By supper time the vegetables and meat were tender, the fire nearly out, but the water still boiling hot. Just at this time he would stir in the corn meal and bran until the mush was as thick as could be conveniently stirred, covered it up tightly and in the morning there was the most delicious breakfast for a flock of hens that could be made.

Put a dash of red paint on the left wing of your turkeys, let your neighbors paint theirs on the right wing or on the shoulder. Have an agreement in the matter and then there will be greater harmony in the fall.

Feather-duster makers buy turkey feathers. The long tail feathers (1) and the wing feathers (2) are the most valuable. The pointers (3) growing on the first joint they do not want. Thrifty housewives cut off and dry the first joint for kitchen use. The long feathers at the root of the tail are also utilized in making dusters. All feathers for sale should be dry-picked and free from soil and blood. To pack these large feathers, put sacking in a box the size of the proposed bundle, lay feathers in flat and straight, press down, draw the sacking over and sew up. Do not put different kinds together. The price of turkey and chicken body feathers is generally low, but by picking over a barrel or box they may be saved without much extra labor. The importance of saving duck and geese feathers need hardly be mentioned.

A most excellent remedy for many sick fowls is composed of a sharp hatchet and a good spade.

A hen hatching ducks is simply brooding over trouble for herself

PLATE XVI.

TOULOUSE AND BROWN CHINA GEESE.

Chapter XVII.

DISEASES AND ENEMIES.

Dampness, filth and roup occupy the same quarters and are fast friends.

A bucket of whitewash is better than a chestful of medicine.

—Tim's Wife.

Many of the ills that poultry flesh is heir to are directly traceable to bad breeding and treatment. In-and-in-breeding is practiced and the law of the survival of the fittest is disregarded until the stock becomes weak and a prey to disease.

Yards and runs occupied for any considerable time become covered with excreta and a breeding ground for all manner of disease germs.

Dampness from leaky roofs or from wet earth floors, and draughts from side cracks, or from overhead ventilation slay their thousands yearly.

A one-sided diet of grain, especially corn, moldy grain or meal, decayed meat or vegetables, filthy water, or the lack of gritty material are fruitful sources of sickness.

In the treatment of sick birds much depends on the nursing and care. It is useless to give medicine unless some honest attempt be made to remove the causes that produce the disturbance. Unless removed the cause will continue to operate and the treatment must be repeated.

It is an excellent plan to have a coop in some secluded place to be used exclusively as a hospital. If cases cannot be promptly treated it is better to use the hatchet at once and bury deeply, or burn the carcasses. This is the proper plan in every case where birds become very ill before they are discovered.

Sick birds should in no case be allowed to run with the flock and to eat and drink with them.

In giving the following remedies I make no pretence to a scientific handling of the subject. Homeopathic remedies are given along with the common drugs. Readers can "pay their money and take their choice."

When the former are used they should be purchased of a homeopathic physician or homeopathic pharmacy. In administering them to fowls able to eat and drink, fifteen or twenty pellets, or five to ten drops of liquid, may be put in a pint of drinking water, or the water may be used to moisten their soft food. If administered to the sick bird directly, a few pellets, four or five, or a tablespoonful of the medicated water may be put down the patient's throat four or five times a day.

FEVERS, from colds, fighting of cocks, etc. Symptoms: unusual heat of body, red face, watery eyes and watery discharge from nostrils.

Give dessertspoonful citrate of magnesia and, as a drink, ten drops of nitre in half a pint of water.

Homeopathic remedy—Aconite, 3, in drinking water.

APOPLEXY AND VERTIGO, from overfeeding or fright. Symptoms: unsteady motion of the head, running around, loss of control of limbs. Give a purgative and bleed from the large veins under wing. Homeopathic—Belladonna, 3. Give a light, non-stimulating diet and keep in a quiet place.

PARALYSIS, from highly seasoned food and over stimulating diet. Symptoms: inability to use the limbs, birds lie helpless on their side. Allopathic treatment—The same as for apoplexy. Homeopathic—Nux vomica, 3.

LEG WEAKNESS occurs in fast-growing young birds, mostly among cockerels. A fowl having this weakness will show it by squatting on the ground frequently and by a tottering walk. When not hereditary it usually arises from a diet that contains too much fat and too little flesh and bone-making material, such as bread, rice, corn and potatoes. To this should be added cut green bone, oats, shorts, bran and clover, green or dry. Give a tonic pill three times a day made of sulphate of iron, 1 grain; strychnine, 1 grain; phosphate of lime, 16 grains; sulphate of quinine, ½ grain. Make into thirty pills. Homeopathic—Calcarea silicata, 6. If occurring in young birds after exposure to dampness or a sudden change to cold weather, give dulcamara, 15.

CANKER OF THE MOUTH AND HEAD. The sores characteristic of this disease are covered with a yellow, cheesy matter which, when it is removed, reveals the raw flesh. Canker will rapidly spread through a flock, as the exudation from the sores is a virulent poison, and well birds are contaminated through the soft feed and drinking water. Sick birds should be separated from the flock and all water and feed vessels disinfected by scalding or coating with lime wash. Apply to sores with a small pippet syringe or dropper the peroxide of hydrogen. When the entire surface is more or less affected, use a sprayer. Where there is much of the cheesy matter formed, first remove it with a large quill before using the peroxide. A simple remedy is an application to the raw flesh of powdered alum, scorched until slightly brown. Homeopathic—Mercurius, vivus or nitric acid internally, with the use of sulphurous acid spray.

SCALY LEG, caused by a microscopic insect burrowing beneath the natural scales of the shank. At first the shanks appear dry, and a fine scale like dandruff forms. Soon the natural scale disappears and gives place to a hard, white scurf. The disease passes from one fowl to another through the medium of nests and perches, and the mother-hen infecting her brood. To prevent its spread, coat perches with kerosene and burn old nesting material and never use sitting hens affected by the disease. To cure, mix ½ ounce flowers of sulphur, ¼ ounce carbolic acid crystals and stir these into 1 pound of melted lard. Apply with an old tooth brush, rubbing in well. Make applications at intervals of a week.

WORMS in the intestines of fowls indicate disturbed digestion. Loss of appetite and lack of thrift are signs of their presence. Give santonin in 2-grain doses six hours apart. A few hours after the second dose give a dessertspoonful of castor oil. Or, put 15 drops of spirits of turpentine in a pint of water and moisten the feed with it. Homeopathic—Cina, 3.

BUMBLE-FOOT, caused by a bruise in flying down from perches or in some similar manner. A small corn appears on the bottom of the foot, which swells and ulcerates and fills with hard, cheesy pus. With a sharp knife make a cross cut and carefully remove all the pus. Wash the cavity with warm water, dip the foot in a solution of one-fourth ounce sulphate of copper to a quart of water and bind up with a rag and place the bird on a bed of dry straw. Before putting on the bandage anoint the wound with the ointment recommended for scaly leg or coat it with iodine.

GAPES, caused by the gape-worm, a parasite that attaches itself to the windpipe, filling it up and causing the bird to gasp for breath. The cut shows the natural size of the parasite as it appears attached to the windpipe. The worm is about three-fourths of an inch long, smooth and red in color. It appears to be forked at one end, but in reality each parasite is two worms, a male and female, firmly joined together; the male is shown at D, and the female, which is the larger of the two, is seen at E. B is a section of the windpipe. This parasite breeds in the common earth worm. Chicks over three months old are seldom affected. If kept off of the ground for two months after hatching, or on perfectly dry soil, or on land where affected chicks have never run, chicks will seldom suffer from the gapes. Old runs and infested soil should have frequent dressings of lime.

In severe cases the worms should be removed. To do this put a few drops of kerosene in a teaspoonful of sweet oil. Strip a soft wing feather of its web to within an inch of the tip as shown in the illustration, dip in the oil, insert feather in windpipe, twirl and withdraw. Very likely some of the parasites and mucus will come with it. The rest will be loosened or killed, and eventually thrown out. It may be necessary to repeat the operation.

To kill the worm in its lodgment, gum camphor in the drinking water or pellets of it as large as a pea forced down the throat is recommended. Turpentine in the soft feed, as advised in the treatment for worms in the intestines, is said to be effective. Pinching the windpipe with the thumb and finger will sometimes loosen the parasite.

When broods are quartered on soil known to be infested, air-slaked lime should be dusted on the floor of the coop, and every other night, for two or three weeks, a little of the same should be dusted in the coop over the hen and her brood. To apply, use a dusting bellows and only a little each time.

CHOLERA is due to a specific germ, or virus, and must not be confounded with common diarrhœa. In genuine cholera digestion is arrested, the crop remains full, there is fever and great thirst. The bird drinks, but refuses food and appears to be in distress. There is a thickening of the blood, which is made evident in the purple color of the comb. The discharges from the kidneys, called the urates, which in health are white, become yellowish, deep yellow, or, in the final stages, a greenish-yellow. The diarrhœa grows more severe as the disease progresses. A fowl generally succumbs in two days. The virus of cholera is not diffusible in the air, but remains in the soil, which becomes infected from the discharges, and in the body and blood of the victims. It may be carried from place to place on the feet of other fowls or animals. Soil may be disinfected by saturating it with a weak solution of sulphuric acid in water. Remove at once all well birds to new and clean quarters and wring the necks of all sick birds and burn their carcasses and disinfect their quarters.

For cases not too far gone to cure give sugar of lead, pulverized opium, gum camphor, of each, 60 grains, powdered capsicum (or fluid extract of capsicum is better, 10 drops), grains, 10. Dissolve the camphor in just enough alcohol that will do so without making it a fluid, then rub up the other ingredients in the same bolus, mix with soft corn meal dough, enough to make it into a mass, then roll it and divide the whole into one hundred and twenty pills. Dose, one to three pills a day for grown chicks or turkey, less to the smaller fry. The birds that are well enough to eat should have sufficient powdered charcoal in their soft feed every other day to color it slightly, and for every twenty fowls five drops of carbolic acid in the hot water with which the feed is moistened.

Homeopathic—Arsenicum, 6, or arsenicum iodatum. As a preventive, use a few drops of camphor in the drinking water.

ROUP. The first symptoms are those of a cold in the head. Later on the watery discharge from the nostrils and eyes thickens and fills the nasal cavities and throat, the head swells and the eyes close up and bulge out. The odor from affected fowls is very offensive. It is contagious by diffusion in the air and by contact with the exudations from sick fowls. To disinfect houses and coops burn sulphur and carbolic acid in them after turning the fowls out and keep closed for an hour or two. Pour a gill of turpentine and a gill of carbolic acid over a peck of lime and let it become slaked, then scatter freely over the interior of houses and coops and about the yards.

For the first stages spray the affected flock while on the roost or in the coop with a mixture of two tablespoonfuls of carbolic acid and a piece of fine salt as big as a walnut in a pint of water. Repeat two or three times a week. Or, if a dry powder is preferred, mix equal parts of sulphur, alum and magnesia and dust this in their nostrils, eyes and throat with a small powder gun. The nasal cavities should be kept open by injecting with a glass syringe or sewing machine oil-can a drop or two of crude petroleum. A little should be introduced also through the slit in the roof of the mouth. Give sick birds a dessertspoonful of castor oil two nights in succession, and feed soft food of bran and corn meal seasoned with red pepper and powdered charcoal. A physician advises the following treatment: hydrastin, 10 grains; sulph. quinine, 10 grains; capsicum, 20 grains. Mixed in a mass with balsam copaiba and made into twenty pills; give one pill morning and night; keep the bird warm and inject a saturated solution of chlorate potash in nostrils and about 20 drops down the throat.

Homeopathic—Aconite, 3, in first stages; mercurius vivus, 6, when the discharge becomes thick; and spongia, 15, when there is a rattling and croupy condition in the throat.

PIP, so-called, is not a disease but only a symptom. The

drying and hardening of the end of the tongue in what is called "pip" is due to breathing through the mouth, which the bird is compelled to do because of the stoppage of the nostrils. By freeing the natural air passages the tongue will resume its normal condition.

DIPHTHERIA is a contagious disease. The first symptoms are those of a common cold and catarrh. The head becomes red and there are signs of fever, then the throat fills up with thick, white mucus and white ulcers appear. The bird looks anxious and stretches its neck and gasps. When it attacks young chicks it is frequently mistaken for gapes. When diphtheria prevails, impregnate the drinking water with camphor, a teaspoonful of the spirits to a gallon of water, and fumigate the house as recommended for roup.

Spray the throat with peroxide of hydrogen or with this formula: 1 ounce glycerine, 5 drops nitric acid, 1 gill water. To treat several birds at once with medicated vapor, take a long box with the lid off, make a partition across and near to one end and cover the bottom with coal ashes. Mix a tablespoonful each of pine tar, turpentine and sulphur, to which add a few drops, or a few crystals, of carbolic acid and a pinch of gum camphor. Heat a brick very hot, put the fowls in the large part and the brick in the other, drop a spoonful of the mixture on the brick and cover lightly to keep the fumes in among the patients. Watch carefully, as one or two minutes may be all they can endure. Repeat in six hours if required.

Homeopathic treatment—Use sulphurous acid spray, and give internally mercurius iodatum, 1, every two hours.

CROP-BOUND. The crop becomes much distended and hard from obstruction of the passage from the crop to the gizzard by something swallowed; generally, it is long, dried grass, a bit of rag or rope. Relief may sometimes be afforded by giving a tablespoonful of sweet oil and then gently kneading the crop with the hand. Give no food, except a little milk, until the crop is emptied. Wet a tablespoonful or more of pulverized charcoal with the milk and force it down the throat. Should the crop not empty itself naturally pluck a few feathers from the upper right side of it and with a sharp knife make a cut about an inch long in the outer skin. Draw this skin a little to one side and cut open the crop. Remove its contents, being careful not to miss the obstruction. Have a needle threaded with white silk ready, and take a stitch or two in the crop skin first, then sew up the outer skin separately. Put the patient in a comfortable coop, and feed sparingly for a week on bran and meal in a moist state, and give but little water.

SOFT OR SWELLED-CROP arises from lack of grit, or from eating soggy and unwholesome food. The distended crop contains water and gas, the bird is feverish and drinks a great deal. By holding it up with its head down the crop will usually empty itself. When this is done give teaspoon doses of charcoal slightly moistened twice at intervals of six hours. Restrict the supply of water and feed chopped onions and soft feed in moderation. Homeopathic—Nux vomica, 3.

EGG-BOUND, DISEASES OF THE OVIDUCT. Overfat hens are often troubled in this way. Forcing hens for egg production will sometimes break down the laying machinery. Give green food, oats, little corn, and no stimulating condiments. Let the diet be plain and cooling in its nature. To relieve hens of eggs broken in the oviduct, anoint the forefinger with sweet oil and deftly insert and draw out the broken parts. When the hen is very fat and the egg is so large it cannot be expelled, the only way to save the hen is to break the egg and remove it as above directed. Homeopathic—Pulsatilla, 3, one day, and calcarea carbonate, 15, the next.

WHITE-COMB OR SCURVY, caused by crowded and filthy quarters and lack of green food. The comb is covered with a white scurf. This condition sometimes extends over the head and down the neck, causing the feathers to fall off.

Change the quarters and diet, give a dose of castor oil and follow this with a half a teaspoonful of sulphur in the soft food daily.

Homeopathic—Sulphur for one day, followed by graphites, 6.

RHEUMATISM AND CRAMP caused by cold and dampness. Chicks reared on bottom-heat brooders are particularly subject to these troubles. Damp earth floors and cement floors in poultry houses produce it in older birds.

Give dry and comfortable quarters, feed little meat, plenty of green food, and soft feed seasoned with red pepper.

Homeopathic—Rhus tox, 3, followed by bryonia, 3.

DIARRHŒA of chicks with clogging of the vent. Homeopathic — Padophyllum, a few drops in the drinking water. Also remove the hardened excretion and anoint the parts. Chamomilla is also useful in this complaint.

DYSENTERY. The symptoms are frequent straining and the passage of urates streaked with blood. Homeopathic—Mercurius corrosivus is indicated.

LOSS OF SIGHT AND WASTING AWAY. Homeopathic—Phosphorous, 6.

FROSTED COMB AND WATTLES. As soon as discovered bathe with compound tincture of benzoin.

FOR LICE on perches, walls and coops, use kerosene or lime wash. To make the lime-wash more effective, pour a little crude carbolic acid on the lime before slaking or mix with plenty of salt.

For use in nests, pour crude carbolic acid on lime and allow it to air-slake. Put one or two handfuls of the carbolized lime dust in the nest box.

Pyrethrum powder, sold as insect powder, is the dry leaves and blossoms of Pyrethrum roseum ground to a fine dust. This kills by contact and is effective for dusting in nests, and through the feathers of birds. It is not poisonous to animal life. Its judicious use in the plumage and nests of sitting hens will insure immunity from lice for the hen and her young brood.

Chicks and poults are often killed by large lice that congregate about the head, throat, vent and wings. To destroy them, soak fish berries (coccolus indicus) in alcohol, take the birds from under the mothers at night and slightly moisten the down of the infested parts with the poison. Kerosene oil, clear, or mixed with sweet oil or lard may be used in the same way if care be taken to use only a little.

RATS, of all vermin, are probably the most destructive because of their number and because they harbor in and around poultry buildings. Cats, terrier dogs, traps and poisons should all be used for their annihilation. Rats have a great liking for ducklings and it is necessary to guard them with special care.

OPOSSUMS will lodge in rail piles during the day-time and raid the coops and houses at night. They kill a few at a time and gnaw the neck and head only. A steel trap set inside at the hole where the animal enters and screened by boards to prevent the fowls from interfering will catch the rascals.

MINKS AND WEASELS will kill a whole coopful in one night. They do not eat but only bleed them in the neck and suck the blood. These vermin live in swampy ravines, whence they come and lodge a few days in brush and rail piles, or along fences while engaged in their work of slaughter. Dogs and traps may be used against them.

FOXES are also night maurauders and their sly games may be foiled by closed coops and houses and a watchful dog.

A good arrangement for trapping all these varmints is shown by the illustration given herewith. It consists of a large box open at both ends having the central part securely enclosed by strong wire netting. A hen and her brood, or a few half-grown chicks or ducklings are put in through the trap-door on top. In both ends steel traps are set and concealed by litter or bits of thin cloth, the traps being securely chained. In the cut the side of coop is left off, to better show its construction.

CATS, generally the innocent-looking pet cat, often acquire a taste for young chickens and will eat two or three daily with great regularity. The best remedy is lead from a shot-gun, or, if the fur is wanted, put pussy and an ounce of chloroform together in a close box.

HAWKS AND CROWS in the vicinity of woods are often troublesome. When they have once caught a chicken at a certain place they will usually come at the same hour the following day. Guineas are useful as alarmists. A shot gun well handled will bring down the enemy. Screens of brush or boards in the yards into which the flocks may run, afford protection. Set a pole with pegs in, to make climbing easy, in the vicinity, nail a small board on top, put a piece of recently killed chicken on it with a steel trap on the chicken and fasten trap with a chain.

Crows catch only small chickens. Suspend in the runs small panes of glass, or pieces of mirror, or bright tin by cords from leaning poles or stakes. These swinging in the wind and glistening in the light are feared by the suspicious thieves. An upright pole may be set in the ground with cross arms and wind-wheel on top, as shown in the illustration.

POT-PIE.

As an evening feed in cold weather nothing is better than whole corn slightly warmed.

Wading in slush is not the kind of exercise that keeps hens healthy and makes them lay in winter.

The public know where Peter Tumbledown's chickens roost by the appearance of his wagon when he drives into town.

An M. D. says that thirty-grain doses of salicylicate of soda will cure fowls of rheumatism.

A large proportion of the substance of an egg is water. Eggs cannot be made out of dry grain and dried grass. Hens that lay in winter have a liberal supply of water from some source.

Sods from a gravelly loam furnish grit, insects, seeds and dried grass. Those who have not tried sods for winter use do not know how valuable they are. Store a big pile in one corner of the hennery.

Moulting fowls require nitrogenous food. Milk, wheat bran and linseed meal, animal meal, cut green bones and the like will furnish it.

The place for unoccupied coops is in a shed or temporary shelter. Clean out and whitewash before putting them away for the season.

Dry feathers in the shade; the sun draws the oil from the stems.

Rotten eggs as nest eggs are an abomination; medicated eggs for keeping away lice are humbugs.

If you have a hen noted for her laying qualities save her eggs and hatch them and raise a few cockerels for next year. This is the way to increase the egg production of your whole flock. Stick a big pin in this item.

A roof that is to be covered with felt of any kind should not be made very steep. If the house is, say, ten feet wide, the roof ought not to have more than twelve or eighteen inches pitch. If two or three feet pitch is given it the wind will get undr the felt and tear it off. We've had experience in the matter and speak "by the book."

To catch a chicken or turkey quick and easy, take a cord and make a slip-noose on one end about twenty inches in diam eter. Lay this on the ground and stand off some distance with the other end in your hand. Throw some corn about the noose and when the right fowl gets his feet within the circle of the cord, pull quick and you have him.

Clover hay is excellent for laying hens. It is rich in the chemical qualities needed in producing eggs. It is also much cheaper than to feed them altogether on grain. Give them grain at night, but in the morning take a pail two-thirds full of fine cut clover and cover with boiling water, cover closely and let it steam until the clover is swelled, then add enough meal, ground oats or bran to make a crumbly mass.

Two handy coops are shown in the illustrations. Figure 1 has ends made of canvas or bagging, and Figure 2 is provided with a sliding false side, which may be drawn to the front by means of the pole, thus bringing the chickens within reach.

FIG. 1.

FIG. 2.

Here are some of the many causes why chicks die in the shell : eggs from immature pullets ; cock too fat ; hens too fat ; hens beginning to moult ; shells of eggs too thick ; cock inactive ; feeding highly-seasoned food ; lack of exercise of hens ; exposing the eggs just when the chicks are coming out ; lack of bulky food for hen ; natural weakness of parents, in-breeding ; lack of vigor in male ; inherent lack of vitality in chicks ; too close and persistent sitting by the hen, thus overheating the eggs ; hens once affected with the roup ; cockerel not matured.

A good condition powder for laying hens or fattening stock : Ground bone, one pound (phosphoric acid and lime) ; ground meat or blood, three pounds (nitrogenous, forming albumen) ; linseed meal, one-half pound (nitrogenous, carbonaceous and laxative, used for regulating the bowels) ; charcoal, one pound (used for promoting digestion and assisting to correct acidity) ; salt, half pound (very necessary. and often neglected) ; ground ginger, two ounces ; red pepper, one tablespoonful ; gentian, one ounce (stimulant and corrective) ; citrate of iron and ammonia, one ounce (an invigorator of the system). A small handful daily to each ten fowls in soft feed.

A good condition powder for sick fowls : gentian, one pound ; red pepper, half ounce ; salt, one ounce ; citrate of iron and ammonia, one ounce ; Peruvian bark, one ounce ; black antimony, one ounce ; charcoal, half a pound. Give a tablespoonful to two hens in the soft feed once a day.

INDEX.